Statistics and Scientific Method

Statistics and Scientific Method

An Introduction for Students and Researchers

PETER J. DIGGLE

and

AMANDA G. CHETWYND

Lancaster University

OXFORD
UNIVERSITY PRESS

Great Clarendon Street, Oxford OX2 6DP

Oxford University Press is a department of the University of Oxford.
It furthers the University's objective of excellence in research, scholarship,
and education by publishing worldwide in

Oxford New York

Auckland Cape Town Dar es Salaam Hong Kong Karachi
Kuala Lumpur Madrid Melbourne Mexico City Nairobi
New Delhi Shanghai Taipei Toronto

With offices in

Argentina Austria Brazil Chile Czech Republic France Greece
Guatemala Hungary Italy Japan Poland Portugal Singapore
South Korea Switzerland Thailand Turkey Ukraine Vietnam

Oxford is a registered trade mark of Oxford University Press
in the UK and in certain other countries

Published in the United States
by Oxford University Press Inc., New York

© Peter J. Diggle and Amanda G. Chetwynd 2011

The moral rights of the authors have been asserted
Database right Oxford University Press (maker)

First published 2011

All rights reserved. No part of this publication may be reproduced,
stored in a retrieval system, or transmitted, in any form or by any means,
without the prior permission in writing of Oxford University Press,
or as expressly permitted by law, or under terms agreed with the appropriate
reprographics rights organization. Enquiries concerning reproduction
outside the scope of the above should be sent to the Rights Department,
Oxford University Press, at the address above

You must not circulate this book in any other binding or cover
and you must impose the same condition on any acquirer

British Library Cataloguing in Publication Data

Data available

Library of Congress Cataloging in Publication Data

Data available

Typeset by SPI Publisher Services, Pondicherry, India
Printed in Great Britain
on acid-free paper by
CPI Antony Rowe, Chippenham, Wiltshire

ISBN 978–0–19–954318–2 (Hbk.)
978–0–19–954319–9 (Pbk.)

1 3 5 7 9 10 8 6 4 2

To Mike, especially for Chapter 8

Acknowledgements

Most of the diagrams in the book were produced using R. For the remainder, we thank the following people and organizations:

Figure 2.1. Institute of Astronomy library, University of Cambridge

Figure 4.1. Professor Andy Cossins, University of Liverpool

Figure 5.1. copyright Rothamsted Research Ltd.

Figure 5.3. copyright photograph by Antony Barrington Brown, reproduced with the permission of the Fisher Memorial Trust

Figure 10.1. Devra Davis (www.environmentalhealthtrust.org)

We have cited original sources of data in the text where possible, but would like here to add our thanks to: Dr Bev Abram for the *Arabadopsis* microarray data; Professor Nick Hewitt for the Bailrigg daily temperature data; Dr Raquel Menezes and Dr José Angel Fernandez for the Galicia lead pollution data; Professor Stephen Senn for the asthma data; CSIRO Centre for Irrigation Research, Griffith, New South Wales, for the glyphosate data; Dr Peter Drew for the survival data on dialysis patients; Professor Tanja Pless-Mulloli for the Newcastle upon Tyne black smoke pollution data.

Preface

Statistics is the science of collecting and interpreting data. This makes it relevant to almost every kind of scientific investigation. In practice, most scientific data involve some degree of imprecision or uncertainty, and one consequence of this is that data from past experiments cannot exactly predict the outcome of a future experiment. Dealing with uncertainty is a cornerstone of the statistical method, and distinguishes it from mathematical method. The mathematical method is deductive: its concern is the logical derivation of consequences from explicitly stated assumptions. Statistical method is inferential: given empirical evidence in the form of data, its goal is to ask what underlying natural laws could have generated the data; and it is the imprecision or uncertainty in the data which makes the inferential process fundamentally different from mathematical deduction.

As a simple example of this distinction, consider a system with a single variable input, x, and a consequential output, y. A scientific theory asserts that the output is a linear function of the input, meaning that experimental values of x and y will obey the mathematical relationship

$$y = a + b \times x$$

for suitable values of two constants, a and b. To establish the correct values of a and b, we need only run the experiment with two different values of the input, x, measure the corresponding values of the output, y, plot the two points (x, y), connect them with a straight line and read off the intercept, a, and slope, b, of the line. If the truth of the assumed mathematical model is in doubt, we need only run the experiment with a third value of the input, measure the corresponding output and add a third point (x, y) to our plot. If the three points lie on a straight line, the model is correct, and conversely. However, if each experimental output is subject to any amount, however small, of unpredictable fluctuation about the underlying straight-line relationship, then logically we can neither determine a and b exactly, nor establish the correctness of the model, however many times we run the experiment. What we can do, and this is the essence of the statistical method, is *estimate* a and b with a degree of uncertainty which diminishes as the number of runs of the experiment increases, and establish the extent to which the postulated model is *reasonably consistent with* the experimental data.

In some areas of science, unpredictable fluctuations in experimental results are a by-product of imperfect experimental technique. This is presumably the thinking behind the physicist Ernest Rutherford's much-quoted claim that 'If your experiment needs statistics, you ought to have done a better experiment.' Perhaps for this reason, in the physical sciences unpredictable fluctuations are often described as 'errors'. In other areas of science, unpredictability is an inherent part of the underlying scientific phenomenon, and need carry no pejorative associations. For example, in medicine different patients show different responses to a given treatment for reasons that cannot be wholly explained by measurable differences amongst them, such as their age, weight or other physiological characteristics. More fundamentally, in biology unpredictability is an inherent property of the process of transmission and recombination of genetic material from parent to offspring, and is essential to Darwinian evolution.

It follows that the key idea in the statistical method is to understand *variation* in data and in particular to understand that some of the variation which we see in experimental results is predictable, or *systematic*, and some unpredictable, or *random*. Most formal treatments of statistics tend, in the authors' opinion, to overemphasize the latter, with a consequential focus on the mathematical theory of probability. This is not to deny that an understanding of probability is of central importance to the statistics discipline, but from the perspective of a student attending a service course in statistics an emphasis on probability can make the subject seem excessively technical, obscuring its relevance to substantive science. Even worse, many service courses in statistics respond to this by omitting the theory and presenting only a set of techniques and formulae, thereby reducing the subject to the status of a recipe book.

Our aim in writing this book is to provide an antidote to technique-oriented service courses in statistics. Instead, we have tried to emphasize statistical concepts, to link statistical method to scientific method, and to show how statistical thinking can benefit every stage of scientific inquiry, from designing an experimental or observational study, through collecting and processing the resulting data, to interpreting the results of the data-processing in their proper scientific context.

Each chapter, except Chapters 1 and 3, begins with a non-technical discussion of a motivating example, whose substantive content is indicated in the chapter subtitle. Our examples are drawn largely from the biological, biomedical and health sciences, because these are the areas of application with which we are most familiar in our own research. We do include some examples from other areas of science, and we hope that students whose specific scientific interests are not included in the subject matter of our examples will be able to appreciate how the underlying statistical concepts are nevertheless relevant, and adaptable, to their own areas of interest.

Our book has its origins in a service course at Lancaster University which we have delivered over a period of several years to an audience of first-year postgraduate students in science and technology. The scientific maturity of students at this level, by comparison with undergraduates, undoubtedly helps our approach to succeed. However, we do not assume any prior knowledge of statistics, nor do we make mathematical demands on our readers beyond a willingness to get to grips with mathematical notation (itself a way of encouraging precision of thought) and an understanding of basic algebra.

Even the simplest of statistical calculations requires the use of a computer; and if tedium is to be avoided, the same applies to graphical presentation of data. Our book does not attempt to teach statistical computation in a systematic way. Many of the exercises could be done using pencil, paper and pocket calculator, although we hope and expect that most readers of the book will use a computer.

We have, however, chosen to present our material in a way that will encourage readers to use the R software environment (see the website www.r-project.org). From our perspectives as teachers and as professional statisticians, R has a number of advantages: its open-source status; the fact that it runs on most platforms, including Windows, Macintosh and Linux operating systems; its power in terms of the range of statistical methods that it offers. Most importantly from a pedagogical perspective, using R encourages the open approach to problems that the book is intended to promote, and discourages the 'which test should I use on these data' kind of closed thinking that we very much want to avoid. Of course, R is not the only software environment which meets these criteria, but it is very widely used in the statistical community and does seem to be here to stay.

We have provided datasets and R scripts (sequences of R commands) that will enable any reader to reproduce every analysis reported in the book. This material is freely available at: www.lancs.ac.uk/staff/diggle/intro-stats-book.

We hope that readers unfamiliar with R will either be able to adapt these datasets and programmes for their own use, or will be stimulated to learn more about the R environment. But we emphasize that the book can be read and used without any knowledge of, or reference to, R whatsoever.

Contents

1	**Introduction**	1
	1.1 Objectives	1
	1.2 Statistics as part of the scientific method	1
	1.3 What is in this book, and how should you use it?	2
2	**Overview: investigating Newton's law**	5
	2.1 Newton's laws of motion	5
	2.2 Defining the question	7
	2.3 Designing the experiment	8
	2.4 Exploring the data	9
	2.5 Modelling the data	10
	2.6 Notational conventions	12
	2.7 Making inferences from data	13
	2.8 What have we learnt so far?	15
3	**Uncertainty: variation, probability and inference**	17
	3.1 Variation	17
	3.2 Probability	21
	3.3 Statistical inference	24
	3.4 The likelihood function: a principled approach to statistical inference	27
	3.5 Further reading	31
4	**Exploratory data analysis: gene expression microarrays**	33
	4.1 Gene expression microarrays	33
	4.2 Displaying single batches of data	36
	4.3 Comparing multiple batches of data	40
	4.4 Displaying relationships between variables	42
	4.5 Customized plots for special data types	45
	4.5.1 Time series	45
	4.5.2 Spatial data	48
	4.5.3 Proportions	49
	4.6 Graphical design	50
	4.7 Numerical summaries	51
	4.7.1 Summation notation	51
	4.7.2 Summarizing single and multiple batches of data	52
	4.7.3 Summarizing relationships	54

5 Experimental design: agricultural field experiments and clinical trials — 57
5.1 Agricultural field experiments — 57
5.2 Randomization — 59
5.3 Blocking — 63
5.4 Factorial experiments — 65
5.5 Clinical trials — 67
5.6 Statistical significance and statistical power — 68
5.7 Observational studies — 70

6 Simple comparative experiments: comparing drug treatments for chronic asthmatics — 71
6.1 Drug treatments for asthma — 71
6.2 Comparing two treatments: parallel group and paired designs — 71
  6.2.1 The parallel group design — 72
  6.2.2 The paired design — 73
6.3 Analysing data from a simple comparative trial — 73
  6.3.1 Paired design — 73
  6.3.2 Parallel group design — 75
6.4 Crossover designs — 77
6.5 Comparing more than two treatments — 78

7 Statistical modelling: the effect of trace pollutants on plant growth — 79
7.1 Pollution and plant growth — 79
7.2 Scientific laws — 80
7.3 Turning a scientific theory into a statistical model: mechanistic and empirical models — 80
7.4 The simple linear model — 83
7.5 Fitting the simple linear model — 86
7.6 Extending the simple linear model — 87
  7.6.1 Transformations — 87
  7.6.2 More than one explanatory variable — 90
  7.6.3 Explanatory variables and factors — 90
  7.6.4 Reanalysis of the asthma trial data — 90
  7.6.5 Comparing more than two treatments — 92
  7.6.6 What do these examples tell us? — 95
  7.6.7 Likelihood-based estimation and testing — 95
  7.6.8 Fitting a model to the glyphosate data — 96
7.7 Checking assumptions — 97
  7.7.1 Residual diagnostics — 99
  7.7.2 Checking the model for the root-length data — 101
7.8 An exponential growth model — 103
7.9 Non-linear models — 107

	7.10 Generalized linear models	108
	7.10.1 The logistic model for binary data	108
	7.10.2 The log-linear model for count data	110
	7.10.3 Fitting generalized linear models	111
	7.11 The statistical modelling cycle: formulate, fit, check, reformulate	112
8	**Survival analysis: living with kidney failure**	**114**
	8.1 Kidney failure	114
	8.2 Estimating a survival curve	115
	8.3 How long do you expect to live?	119
	8.4 Regression analysis for survival data: proportional hazards	122
	8.5 Analysis of the kidney failure data	123
	8.6 Discussion and further reading	126
9	**Time series analysis: predicting fluctuations in daily maximum temperatures**	**127**
	9.1 Weather forecasting	127
	9.2 Why do time series data need special treatment?	127
	9.3 Trend and seasonal variation	128
	9.4 Autocorrelation: what is it and why does it matter?	130
	9.5 Prediction	133
	9.6 Discussion and further reading	138
10	**Spatial statistics: monitoring air pollution**	**141**
	10.1 Air pollution	141
	10.2 Spatial variation	144
	10.3 Exploring spatial variation: the spatial correlogram	144
	10.4 Exploring spatial correlation: the variogram	146
	10.5 A case-study in spatial prediction: mapping lead pollution in Galicia	147
	10.5.1 Galicia lead pollution data	147
	10.5.2 Calculating the variogram	147
	10.5.3 Mapping the Galicia lead pollution data	148
	10.6 Further reading	156
	Appendix: The R computing environment	**158**
	A.1 Background material	158
	A.2 Installing R	159
	A.3 An example of an R session	160
	References	**163**
	Index	**167**

When you can measure what you are speaking of and express it in numbers, you know that on which you are discoursing. But when you cannot measure it and express it in numbers, your knowledge is of a very meagre and unsatisfactory kind.

Lord Kelvin (Sir William Thomson)

Having numbers is one thing, having them understood and correctly interpreted is quite another.

T. N. Goh

1
Introduction

1.1 Objectives

Our objectives in writing this book are:

- to provide students with a basic understanding of the role that statistics can play in scientific research;
- to introduce students to the core ideas in experimental design, statistical inference and statistical modelling;
- to prepare students for further reading, or for more specialized courses appropriate to their particular areas of research.

Throughout, we emphasize the underlying concepts rather than the technical details. Knowing how to translate a scientific question into a statistical one, and to interpret the result of a statistical analysis, are more important than knowing how to carry out the computations involved in the analysis. Machines can do the computations for us.

An emphasis on concepts rather than on specific techniques distinguishes the book from most introductory statistical texts. Readers who need more technical detail on specific techniques have many excellent books available to them, usually with titles like *Statistics for 'X'*, where 'X' might be biology, sociology, ... Our own favourite amongst these is Altman (1991), whose focus is on medical applications. A corollary to our focus on concepts is that we cover several topics not normally found in introductory texts, including the analysis of time series and spatial data, and give suggestions for further reading in these relatively specialized areas.

1.2 Statistics as part of the scientific method

The goal of science is to understand nature. The two pillars of the scientific method are theory and observation. A scientific theory predicts how a natural process should behave. Observation, whether through a controlled experiment or direct observation of the natural world, can tell us whether the theory is correct. Or, more accurately, it can tell us whether the theory is not correct. A scientific theory cannot be proved in the rigorous sense

2   INTRODUCTION

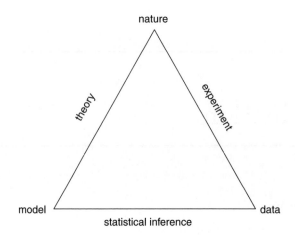

Fig. 1.1. Scientific method and statistical method.

of a mathematical theorem. But it can be falsified, meaning that we can conceive of an experimental or observational study that would show the theory to be false. Well-known examples from the early history of science include the falsification of the pre-Copernican theory that the Sun rotated around a stationary planet Earth, or that the Earth itself was flat. Similarly, the subservience of theory to observation is important. The American physicist Richard Feynman memorably said that 'theory' was just a fancy name for a guess. If observation is inconsistent with theory then the theory, however elegant, has to go. Nature cannot be fooled.

The inferential character of statistical method is in complete harmony with this view of science. Contrary to popular belief, you cannot prove anything with statistics. What you can do is measure to what extent empirical evidence, i.e., data, is or is not consistent with any given theory.

This view of the role of statistics within scientific method is summarized in Figure 1.1. Nature sits at the apex of the triangle; a scientific *theory* provides a model of how nature should behave; a scientific *experiment* generates data that show how nature actually behaves. *Statistical inference* is the bridge between model and data.

Another important role for statistics that is not shown in Figure 1.1 is to help in the design of a scientific investigation so as to make the degree of uncertainty in the results of the investigation as small as possible.

1.3   What is in this book, and how should you use it?

Chapter 2 uses a simple undergraduate physics experiment to illustrate how statistical ideas can be used to advantage in all stages of a scientific investigation. It explains how the key statistical concepts of *design*, *inference* and *modelling* contribute to the process of scientific inquiry.

Chapter 3 discusses the idea of uncertainty and introduces probability as a measure of uncertainty. These concepts underpin the whole of statistics. The less theoretically inclined reader may find some of the ideas difficult to grasp on first reading, but if this applies to you we hope that you will persevere, perhaps after reading the later chapters that are more strongly rooted in scientific applications.

Chapter 4 describes some graphical and numerical techniques that can be used for exploring a set of data without any specific scientific goal in mind. This process, called exploratory data analysis, cannot lead to definitive conclusions but is an essential first step in understanding the patterns of variation in a dataset. At the very least, exploratory data analysis can detect gross errors in the data. More subtly, it can help to suggest suitable models that can be used to answer specific questions. This is especially the case when the underlying scientific theory is only incompletely understood. For example, we may know that toxic emissions from industrial activity harm plant life, but not know the form of the dose-response relationship.

Chapter 5 discusses the statistical approach to experimental design, with an emphasis on the core concepts of *randomization* and *blocking*. We focus initially on agricultural field experiments, but also describe the basic elements of a type of medical research investigation known as a *clinical trial*.

Chapter 6 introduces some simple statistical methods that feature in most introductory courses and textbooks. Our hope is that you will see these methods as logical consequences of the general ideas presented in earlier chapters, rather than as arbitrary formulae plucked from the air.

Chapter 7 discusses statistical modelling. A statistical model is a way of describing mathematically the scientific process that has generated a set of experimental data, in a way that recognizes both predictable and unpredictable variation in the results of the experiment. Historically, many of the most widely used statistical methods were tailored to the analysis of particular types of experiment. Statistical modelling puts all of these methods, and more, into a unified framework. Because of this breadth of scope, Chapter 7 is the longest in the book. If you wish, you can skip Sections 7.9 and 7.10 on a first reading.

Chapters 8, 9 and 10 introduce a range of more specialized statistical methods relating to variation in time or space. In Chapter 8 we consider data in the form of lifetimes, whose distinctive feature is that the data are often *censored*. For example, in studying the survival times of patients being treated for a potentially fatal illness, some patients will die within the study period, whilst others will not: their survival times are said to be *censored* at the study end-time. Chapter 9 concerns data in the form of a time series. A *time series* is a sequence of measurements made at equally spaced times, for example a set of daily recorded levels of air pollution at a specific location. Chapter 10 concerns *spatial* data, for example recorded levels of air pollution at many locations.

Each of Chapters 2 and 3 could be read as a self-contained essay. Chapters 4 to 6 and the early parts of Chapter 7 contain the material needed to design and analyse simple experiments. Chapters 8 to 10 can be read in any order.

You can work through all of the material without doing any statistical calculations yourself. You should then be better able to have an effective dialogue with a statistician kind enough to do the hands-on analysis for you. However, statisticians (although invariably kind) are also scarce, and we would recommend that you learn to reproduce the analyses presented in each chapter, either using the R code and data provided on the book's website, www.lancs.ac.uk/staff/diggle/intro-stats-book, or by using whatever other statistical software you prefer.

We have chosen not to give exercises at the end of each chapter. In our view, the best way to be sure that you have understood the material is to explore different ways of analysing the datasets discussed in the book or, even better, to try using the methods on your own data. We hope that the material on the book's website will help you to build your confidence as you try different things. And we would encourage you to ask yourself at every stage how each statistical analysis that you perform on your own data contributes to your *scientific* understanding.

Even if you do all of the above, you may still find that your investigation needs more sophisticated statistical design and/or analysis than the book covers – it is an introduction, not a comprehensive manual. But we hope that it will give you the knowledge and confidence that you need either to tackle more advanced material yourself, or to engage productively with your statistical colleagues. In our view, statisticians have at least as much to gain from collaborating with scientists as scientists have to gain from collaborating with statisticians.

2
Overview: investigating Newton's law

2.1 Newton's laws of motion

Sir Isaac Newton (1642–1727) was one of the great scientific thinkers of his, or any other, age. Amongst his many achievements, he discovered the 'laws of motion' which form the foundation of classical, or 'Newtonian' mechanics. Discoveries by twentieth-century physicists, amongst whom the best known to the general public is Albert Einstein (1879–1955), reveal Newton's laws to be, strictly, approximations, but approximations that for all practical purposes give excellent descriptions of the everyday behaviour of objects moving and interacting with each other.

One of Newton's laws predicts that an object falling towards the ground will experience a constant acceleration, whose numerical value is often represented by the symbol g, for 'gravity'. It follows using the mathematical tools of the calculus that if an object is dropped from a resting position above the ground, the distance d that it will fall in time t is given by the formula

$$d = \frac{1}{2}gt^2. \tag{2.1}$$

Like any scientific law, (2.1) will only hold under 'ideal conditions'. The statistical method often comes into its own when dealing with experimental data that, for practical reasons, have to be collected in less-than-ideal conditions, leading to discrepancies between the data and the laws which purport to generate them. Some years ago, during a short course on statistics for physics students, we devised a simple lab experiment to illustrate this in the context of Newton's law (2.1). We shall use this same experiment, and the data obtained from it, to motivate an overview of the role played by the statistical method in scientific investigation. We acknowledge at the outset that the experiment used deliberately crude apparatus in order to serve its educational purpose. It would have been easy to set up the experiment in such a way that the data would have obeyed (2.1) more or

Fig. 2.1. Sir Isaac Newton.

less exactly – indeed, when we first asked for the apparatus to be set up for a demonstration, the lab technician did just this, and clearly doubted our sanity when we said that we wanted something *less* precise.

Incidentally, Newton was also one of two apparently independent discoverers of the calculus, the other being the German mathematician Gottfried Wilhelm von Leibnitz (1646–1716); see, for example, Hall (1980). Accounts of Newton's life and scientific achievements include Iliffe (2007) and Gleick (2003). Figure 2.1 shows one of several well-known portraits of Newton.

Returning to our lab experiment, the set-up is illustrated in Figure 2.2. A steel ball-bearing is held by an electromagnet at a vertical distance d above the bench. The value of d can be adjusted using a screw-clamp and measured against a millimetre scale. A two-way switch is wired to the electromagnet and to a stopwatch so that in one position the electromagnet is *on* and the stopwatch is *off*, and in the other the electromagnet is *off* and the stopwatch is *on*. A single run of the experiment consists of setting a value for d, releasing the ball-bearing by throwing the switch, and recording by a second throw of the switch the time elapsed until the ball-bearing hits the bench.

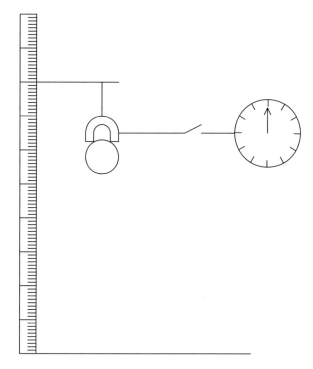

Fig. 2.2. Schematic representation of the apparatus for the lab experiment.

2.2 Defining the question

Most introductory statistics books focus on methods for analysing data. In this book, we postpone our discussion of how to analyse data until we have considered two more basic questions: *why* do we want to analyse the data? and *what* data do we want to collect in the first place? The *why* question is for the scientist to answer. The question is relevant to a statistics book because the scientific purpose of an investigation, be it a laboratory experiment or an observational study of a natural system, will influence what kind of data are to be collected and therefore the statistical tools that will be needed to analyse the resulting data. The data must be logically capable of answering the scientific question but, as we discuss in Section 2.3, they should also do so in as efficient a manner as possible.

In our simple lab experiment, the task we set our physics students was: suppose that the value of the physical constant g in (2.1) were unknown; using a fixed number of runs of the basic experiment, estimate the value of g.

2.3 Designing the experiment

The role of the statistical method in experimental design is, firstly to ensure *validity*, secondly to maximize *efficiency*. By validity, we mean that the experiment is capable of answering the question of scientific interest; by efficiency, we mean that the answer is as precise as possible. In simple experiments, ensuring validity is usually straightforward, ensuring efficiency is less so. The key to both lies in understanding why experimental results might vary between runs, i.e., in identifying all possible *sources of variation*.

For our simple laboratory experiment, the most obvious source of variation is the chosen value for the vertical distance d. Larger values of d will clearly tend to deliver longer measured times t. We call this a source of *systematic* variation, meaning that it is under the control of the experimenter. A second source of variation is the experimenter's reaction-time, which is a consequence of the crude way in which the experiment was set up so that the measured time involves the student responding to the sound of the ball-bearing hitting the bench. Conceivably, the student could anticipate the ball-bearing hitting the bench, but in either event the discrepancy between the actual and the recorded time t will vary unpredictably between repeated runs of the experiment. We call variation of this kind *random variation*.

Sources of variation may also be classified as *of interest* or *extraneous*. In our experiment, variation due to changing the value of d is of interest, because it relates to the scientific question, whereas variation due to reaction time is extraneous. But to a physiologist, the reverse might be true. In other words, the distinction between systematic and random variation is an inherent property of an experiment, whereas the distinction between variation of interest and extraneous variation is context-dependent.

An experimental design should aim to eliminate extraneous sources of variation whenever possible. For example, in our experiment the same student recorded all of the measured times, thereby eliminating variation between the average reaction times of different students. When it is not possible to eliminate extraneous sources of variation, we use randomization to protect the validity of the experiment. In some circumstances, the order in which runs of the experiment are performed may affect the resulting measurement. This could apply in our experiment if the student's average reaction time improved with practice. If this were the case, and the runs were performed in order of increasing values of d, then the variation of interest, namely how the measured time varies with d, would be mixed up with the extraneous variation due to the practice effect. This is called *confounding*, and is clearly undesirable. We eliminate it by first choosing our values of d, then performing the runs of the experiment in a random order.

Table 2.1. Data from the lab experiment.

t (sec)	d (cm)	t (sec)	d (cm)	t (sec)	d (cm)
0.241	10	0.358	40	0.460	70
0.249	10	0.395	45	0.485	75
0.285	15	0.435	50	0.508	80
0.291	20	0.412	50	0.516	85
0.327	25	0.451	55	0.524	90
0.329	30	0.444	60	0.545	90
0.334	30	0.461	65		
0.365	35	0.481	70		

Randomization is fundamental to good statistical design, but this should not be taken as an excuse not to think of other ways in which extraneous variation can be eliminated. Typically, randomization on its own can ensure validity of an experiment, but does not deliver efficiency.

Table 2.1 shows the design used by one of our students and the results obtained by them. Their results are presented in order of increasing d, rather than in time order; note that they include some pairs of repeated measurements at the same value of d, called *replicates*. Replicates allow us to assess the random variation in the data separately from the systematic variation. This is especially important when we do not have a well-established theoretical law, like (2.1), to describe the anticipated pattern of the systematic variation in the data.

2.4 Exploring the data

Once an experiment has been conducted, its recorded results constitute the *data*, which are the raw material from which we now seek to answer the original scientific question. Data analysis is usually conducted in two stages. The first, informal stage is called *exploratory* data analysis. Its role is to describe the general pattern of variation in the data and to look for unexpected features that might point to problems in the experiment, unanticipated sources of variation or simple recording errors in the data. The second, more formal phase, which we discuss in Section 2.7, provides the answer to the original question, and is sometimes called *confirmatory* data analysis.

Graphical methods feature prominently in exploratory data analysis. For our simple experiment, the most obvious form of graphical presentation is a plot of t against d, called a *scatterplot*. Figure 2.3 gives an example. Good graphical design can materially improve the interpretability of plots like Figure 2.3, and we shall discuss this in more detail in Chapter 4. For

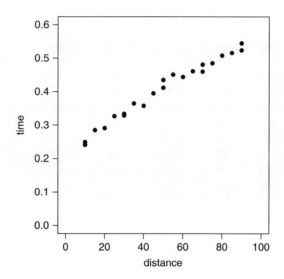

Fig. 2.3. A scatterplot of time against vertical distance for the data from the lab experiment.

the time being, we comment only that a general statistical convention in drawing a scatterplot is to plot the *output* from each run of the experiment, here the recorded time t, on the vertical axis, and the corresponding *input*, here the distance d, on the horizontal axis. At first sight, this contravenes the mathematical convention that in a graph of a function such as (2.1) the horizontal axis should represent the argument of the function, here t, and the vertical axis the corresponding value of the function, here $d = \frac{1}{2}gt^2$. We can reconcile this by re-expressing the law as

$$t = \sqrt{2d/g}. \tag{2.2}$$

Why does this apparently trivial re-expression matter? It matters because the random variation in the experimental data affects t, not d. Hence, the points in our graph (Figure 2.3) can be thought of as randomly perturbed versions of the points on the curve prescribed by (2.2). Or can they?

2.5 Modelling the data

The word *model* is very widely used in science to mean a mathematical representation of a natural system. Models are rarely exact representations of nature, although some, like Newton's laws, are pretty good approximations. Indeed, the best models are those that are informed by well-founded scientific theory. Models of this kind are sometimes called *mechanistic* models, in contrast to *empirical* models that seek only to describe the patterns observed in experimental data.

In the absence of Newton's laws, we might be tempted to describe the pattern in Figure 2.3 by a straight-line relationship between the input d and the output t. This would allow us to quantify the *rate* at which t increases with d as the slope of the best-fitting line. But closer inspection would reveal a discrepancy between the straight-line model and the data; there is a curvature in the relationship for which, in this experiment, Newton's law provides a ready explanation. However, suppose that in equation (2.2) we redefine an input variable $x = \sqrt{d}$, a constant $\beta = \sqrt{2/g}$ and relabel t as y to emphasize that it is an output rather than an input. Then, (2.2) becomes

$$y = \beta x, \qquad (2.3)$$

which *is* the equation of a straight-line relationship between the input x and the output y, with slope β.

Equation (2.3) invites us to draw a scatterplot of y against x, which we show here as Figure 2.4. This does indeed show a linear relationship, but there is still a discrepancy between theory and data; equation (2.3) suggests that the best-fitting straight line should pass through the origin, whereas extrapolation of a line fitted by eye to the data in Figure 2.4 clearly suggests a positive value of y when $x = 0$. This positive value, α say, corresponds to the student's average reaction time, and to accommodate this we extend

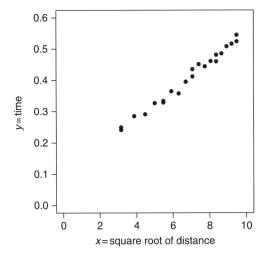

Fig. 2.4. A scatterplot of time against the square root of vertical distance for the data from the lab experiment.

our model to

$$y = \alpha + \beta x. \tag{2.4}$$

This extended model still does not 'fit' the data perfectly, because the data points do not follow exactly a straight line, or indeed any other smooth curve. Reaction time again provides the explanation. As is well known, repeated measurements of a person's reaction time show unpredictable, or *random* variation about an average value. In (2.4), the constant α represents the average reaction time, but there is nothing to represent the random variation in reaction time. Hence, our final model for this simple experiment is

$$y = \alpha + \beta x + Z, \tag{2.5}$$

where now Z is a *random variable*, a quantity which varies unpredictably between runs of the basic experiment.

To summarize: Newton's law tells us to expect a straight line, or *linear* relationship between x and y, of the form $y = \beta x$; basic knowledge of physiology tells us to add an average reaction time α, to give $y = \alpha + \beta x$; the evidence from the data confirms that we need to add a random variable Z, giving the final form of the model as $y = \alpha + \beta x + Z$.

Now, how can we use the model to answer the question?

2.6 Notational conventions

We shall generally use letters near the end of the alphabet $(x, y, \ldots)$ to represent *variables*, and letters near the beginning of the alphabet $(c, g, \ldots)$ to represent known *constants*. However, we may disobey our own rules, for example in using d as a variable. Notation is intended to be helpful, and rules are there to be broken when there is a good reason for doing so; here, the mnemonic value of d for 'distance' justifies our overriding the 'end of the alphabet' convention for variable names.

Statisticians use upper-case letters to denote *random* variables and lower-case letters to denote *non-random* variables, and we will adopt this convention from now on. Hence, a more correct version of equation (2.5) would be

$$Y = \alpha + \beta x + Z. \tag{2.6}$$

This is because x, the square root of the distance, varies non-randomly by design, whereas Z, the variation in the experimenter's reaction time, varies randomly and so, as a consequence, does Y, the time recorded by the experimenter.

The distinction between non-random and random variables leads us to another convention, and initially a confusing one, in which we use lower- and upper-case versions of the same letter to mean subtly different things. In using a lower-case y in equation (2.5), we are thinking of y as one of the numbers in the relevant column of Table 2.1, whereas in (2.6), y has become Y to indicate a model for a generic run of the experiment, in which Y is a random variable because it includes the random variable Z.

We use Greek letters $(\alpha, \beta, \ldots)$ to denote constants whose values are unknown. We call unknown constants *parameters*, a word of Greek origin whose literal translation is 'beyond measurement'. Statisticians always use 'parameter' in this specific sense. Parameters are the embodiment of the scientific questions posed by the data, and as such always occupy centre stage in confirmatory statistical analysis.

2.7 Making inferences from data

Inference is the formal process by which statisticians reach conclusions from data. The inferential paradigm is that the data are but one of infinitely many hypothetical datasets that could be obtained by repeating the experiment under identical conditions. Since such repetitions will produce different data (different versions of Table 2.1), they will also lead to different answers, all equally valid. But the underlying truth, the state of nature, remains the same. For this reason, the answer that the statistical method gives to any scientific question must be couched in terms of uncertainty. This is the kind of thing that gets statistics a bad name, but it is in fact its fundamental strength. The answer to any non-trivial scientific question *is* uncertain. We can reduce the uncertainty by doing more and better experiments, to the point where the uncertainty becomes negligible and we are prepared to behave as if we have an exact answer, but until we reach this happy state of affairs, it is better to acknowledge and quantify our uncertainty rather than to pretend it does not exist.

More specifically, *statistical inferences* fall broadly under three headings: parameter estimation; hypothesis testing; and prediction.

Parameter estimation consists of using the data to make a best guess at the true value of any parameter in a model. Our model (2.5) has three parameters. Two of these, α and β, describe how the average recorded time varies according to the conditions of the experiment, i.e., the chosen value of x. The third, which does not appear explicitly in equation (2.5), relates to the amount of variability on the random term Z. The mathematical interpretation of α and β is as the intercept and slope of the line describing the relationship between x and y. Their scientific interpretation is that α is the average reaction time, whilst β is related to the physical constant, g, by the equation $\beta = \sqrt{2/g}$, or equivalently, $g = 2/\beta^2$. Because of this exact relationship, an estimate of β implies an estimate of g.

A *point estimate* is a single value. For example, fitting a straight line by eye to Figure 2.4 would give point estimates somewhere around $\hat{\alpha} = 0.1$ and $\hat{\beta} = 0.05$. We use a circumflex, or 'hat' above the name of the parameter to indicate an estimate, rather than its true, unknown value. Another person fitting the line by eye would get slightly different estimates, and an objective method of estimation would be preferable. Not only does an objective method yield a unique answer for a given dataset, it allows us to use statistical theory to say by how much we would expect the answer to vary over repetitions of the experiment. This leads to an *interval estimate* of a parameter, a range of values which in some reasonable sense is 'likely' to contain the true value of the parameter in question. A widely accepted statistical method of estimation for our model and data, which we will discuss in detail in Chapters 3 and 7, gives interval estimates for α and β with the property that each has a 95% chance of including the true value of the parameter. This gives the interval estimate for α as the range from 0.0760 to 0.1120, which we write as $\hat{\alpha} = (0.0760, 0.1120)$. In the same notation, the interval estimate for β is $\hat{\beta} = (0.0431, 0.0482)$. Using the relationship that $g = 2/\beta^2$, we can convert the second of these interval estimates to an interval estimate for g that again has a 95% chance of including the true value of g; we calculate the lower and upper limits as $2/0.0482^2 = 860.9$ and $2/0.0431^2 = 1076.7$, to give the interval estimate $\hat{g} = (860.9, 1076.7)$. This interval is the statistical answer to the question, and its width is a measure of how precise is our experiment.

Parameter estimation seeks to establish what range of values might reasonably be assigned to an unknown parameter. *Hypothesis testing* asks, more specifically, whether a *particular* value is reasonably consistent with the data. For example, an assertion that $g = 981 \, \text{cm/sec}^2$ is a *hypothesis*, and a *test* of this hypothesis leads either to its *acceptance* or *rejection* according to whether the hypothesized value is or is not reasonably consistent with the data. This begs the question of what we mean by 'reasonably consistent'. One simple answer is to accept any hypothesized value which falls within the corresponding interval estimate. Amongst other things, this has the desirable consequence of emphasizing that acceptance is not proof, since infinitely many different hypothesized values are thereby accepted, but only one can be true. It follows that formally testing a single hypothesis, for example, $g = 981 \, \text{cm/sec}^2$, only makes sense if the number 981 has some special status that sets it apart from all other values. This might be the case if an experiment is intended to investigate whether a scientific theory known to hold under certain conditions does or does not continue to hold under new, previously unexplored circumstances. In such cases, we can consider a model in which a parameter θ should, according to currently accepted theory, take a particular value, say zero, and test the hypothesis that $\theta = 0$. The formal outcome of a hypothesis test is to accept or reject a hypothesis. It follows that the outcome may be wrong in either of two ways: we may reject a hypothesis that is, in fact, true; or we may accept a hypothesis that is false. A 'good' test is therefore one that makes both of these unlikely. We shall continue this discussion in Chapter 5.

Hypothesis testing played a major role in the development of statistical theory and methods during the first half of the twentieth century, and continues to feature prominently in many elementary statistics courses, and in books aimed at non-statistical audiences. However, in the absence of any prior theory which gives a special status to a particular value of the parameter in question, hypothesis testing is of doubtful scientific value and should be avoided in favour of parameter estimation.

Both parameter estimation and hypothesis testing are concerned with the search for scientific truth. *Prediction* is concerned with establishing what behaviour might be observed as a consequence of that truth. Prediction becomes a statistical exercise when the truth is known only within non-negligible limits of uncertainty. Suppose, in our simple experiment, we wish to predict what time might be recorded in a future run of the experiment with $d = 200$. The existing data are of no direct help, because $d = 200$ falls outside their range. But if we believe our model (2.5), we can plug in point estimates for α and β and simply extrapolate the resulting line to $x = \sqrt{200} \approx 14.14$. Using the mid points of the intervals quoted earlier for $\hat{\alpha}$ and $\hat{\beta}$, the resulting extrapolation is $\hat{Y} = 0.0940 + 0.0457 \times 14.14 = 0.7521$. But since our point estimates are uncertain, so must be our prediction. If we make due allowance for this, it turns out that a reasonable prediction interval is $\hat{Y} = (0.7097, 0.7704)$. We can also make a distinction between predicting the *actual* value of Y we would observe in a single run with $x = 14.14$ and the *average* value of Y over repeated runs with $x = 14.14$. The point prediction is the same for both, but the prediction interval for a future observation is wider than the interval for the average, because of the additional uncertainty that attaches to any future value of Z. This additional uncertainty does not apply to the prediction interval for the average, because by definition the average value of Z is zero. The 95% confidence interval for the average value of Y when $x = 14.14$ is $(0.7207, 0.7594)$, slightly narrower than the corresponding prediction interval $\hat{Y} = (0.7097, 0.7704)$.

2.8 What have we learnt so far?

Statistical thinking can contribute to every stage of scientific inquiry: in the *design* of an experiment before it is conducted; in the preliminary, or *exploratory analysis* of the data that result from the experiment; in formulating a *model*; and in drawing valid scientific conclusions, or *inferences* from the experiment. In reality, the progress from one stage to another may be far from smooth. An ideal design may be infeasible or too expensive to use in practice; exploratory analysis may reveal unforeseen features of the data, prompting a reassessment of the scientific objectives. A model that seems plausible beforehand may prove to be inconsistent with the data – in this respect, we emphasize that the most successful models are often those that strike a balance between, on the one hand purely empirical models that ignore basic scientific knowledge, and on the other highly elaborate mechanistic models that seek to describe natural processes in such fine detail that the available data cannot hope to validate the modelling assumptions. The middle ground has been called 'data-based mechanistic' modelling by Young and Beven (1994).

Fundamentally, the role of the statistical method is to deal appropriately with unpredictable *variation* in experimental results. If the result of an experiment can be reproduced *exactly* by routine repetition, then there is no place for statistics in interpreting the experimental data. This is sometimes incorrectly taken to imply that statisticians are more interested in the 'noise' than in the 'signal' associated with a set of data. A more accurate statement would be that statisticians share scientists' interest in the signal, but recognize that noise may be hard to eliminate altogether, and should be recognized in order to ensure valid scientific conclusions. We once gave a talk in which we introduced a relatively complex statistical model to analyse data that had previously been analysed by much simpler methods. In the subsequent discussion we were accused of 'muddying the water', to which our response was that we were simply acknowledging that the water was, indeed, muddy.

Inference is simply a technical term for the process by which statisticians analyse data in such a way that they can give an honest assessment of the precision, or degree of uncertainty, that must be attached to the conclusions. Often, inferences are disappointingly imprecise, but it is surely better to recognize this than to attach false precision to what may then turn out to be irreproducible results.

3
Uncertainty: variation, probability and inference

3.1 Variation

Whenever a scientist deliberately changes the conditions of an experiment, they would not be surprised to see a consequential change in the results, however expressed. An environmental scientist may grow plants in controlled environments to which different amounts of pollutant have been introduced, so as to understand the adverse effects of pollution on plant growth; a pharmaceutical company developing a new drug may test different doses to investigate how dose changes efficacy and/or the rate of occurrence of undesirable side effects; an engineer may experiment with using different components in a complex piece of equipment to discover how different combinations of components affect the overall performance of the equipment. This kind of *variation* in experimental results is called *systematic* and the variables that induce it are called *design variables* or *factors*; a common convention is to use the first term for quantitative variables (e.g., concentration of pollutant, dose of drug) and the second for qualitative variables (e.g., selection of components of type A, B, C, etc.).

Even when all anticipated sources of systematic variation are held constant, repetition of an experiment under apparently identical conditions often leads to different results. Variation of this kind is called *random*, or *stochastic*. The latter term is really just a fancy version of the former, but is a helpful reminder that randomness can take many forms beyond the everyday notion of random as a fair draw in a lottery. The formal definition of *stochastic* is *governed by the laws of probability* (see Section 3.2 below). Informally, we use probability as a mathematical model of uncertainty in the following sense: stochastic variation in the result of an experiment implies that the result cannot be predicted exactly. In some branches of the physical sciences, the underlying scientific laws that determine the result of an experiment are sufficiently well understood, and the experimental

18   UNCERTAINTY

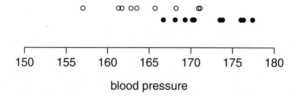

Fig. 3.1. Synthetic data on blood-pressure measurements for 20 subjects, 10 of whom are assigned drug A (solid dots), the remainder drug B (open circles).

conditions sufficiently well controlled, that stochastic variation is negligible. Eliminating stochastic variation is highly desirable, but not always achievable, and certainly not in many branches of environmental science, biology, medicine and engineering.

A third kind of variation arises when variables whose values are not specified in advance are measured in the course of the experiment, even though they are not of direct interest. The data for the following example are synthetic, but show realistic patterns of variation. Suppose that a clinician wishes to compare the efficacy of two different anti-hypertensive drugs, A and B, say. They recruit 20 of their patients, all of whom are being treated for hypertension, and give 10 of them a two-week course of drug A, the remaining 10 a two-week course of drug B (the important question of *how* to choose which 10 patients should be given drug A will be discussed in Chapter 5). Two weeks later, the clinician measures the (systolic) blood pressure of all 20 patients and obtains the following results:

Drug A: 177, 169, 170, 167, 176, 174, 170, 174, 176, 168
Drug B: 169, 162, 157, 164, 164, 163, 161, 171, 171, 166

These data are shown graphically in Figure 3.1. Each value in the data is represented by a dot (solid for results from patients given drug A, open for drug B), whose position has been 'jittered' horizontally to avoid over-plotting.

Drug B appears to give the better result overall, but the measured values of blood pressure show substantial variation between patients. Consequently, the correct interpretation of the data is not clear-cut. A second clinician remarks that in their experience, overweight people tend also to suffer more seriously from hypertension, and suggests calculating the body mass index (BMI) of each patient. This is easily done from medical records, and leads to the data shown in Table 3.1.

Figure 3.2 plots the measured blood-pressure values against BMI, using different plotting symbols to distinguish between drug A and drug B. The diagram shows two things. Firstly, there is a relationship between blood pressure and BMI: people with relatively high BMI do tend to have relatively high blood pressure. Secondly, the fact that the solid dots

Table 3.1. Synthetic data showing measured blood pressure (BP) and body mass index (BMI) for 20 hypertensive patients, 10 of whom were given drug A, the remainder drug B.

Drug A		Drug B	
BP	BMI	BP	BMI
177	27.9	169	27.4
169	24.2	162	23.9
170	25.2	157	25.0
167	26.0	164	26.2
176	26.4	164	26.3
174	26.8	163	27.1
170	26.9	161	27.2
174	28.1	171	27.6
176	29.0	171	28.2
168	25.3	166	26.0

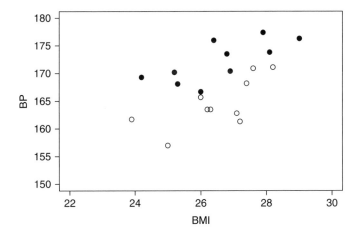

Fig. 3.2. Systolic blood pressure and body mass index for 20 subjects, 10 of whom are assigned drug A (solid dots), the remainder drug B (open circles).

and open circles occupy different parts of the diagram, with most of the open circles lying towards the lower-right part, suggests that once the relationship between BMI and blood pressure is taken into account, drug B is superior to drug A. Put another way, Figure 3.2 suggests that if it had been possible to compare the two drugs on people whose BMI was the same, drug B would have produced consistently lower values of blood pressure than drug A.

In this example, the BMI variable was not of direct scientific interest, but its inclusion in the data enabled a more incisive analysis and a conclusion that drug B is more effective than drug A in treating hypertension. Variables of this kind are called *explanatory variables*, or *covariates*. We favour the first of these, because it aptly describes their role in explaining what would otherwise be treated as stochastic variation. From a mathematical perspective, explanatory variables and design variables are equivalent; both are treated as if fixed by the experimenter and the remaining variation in the data regarded as stochastic. From a scientific perspective, the difference between design variables and explanatory variables is that the experimenter can control the value of a design variable, but not of an explanatory variable. The distinction can be important. For example, one universal principle of research involving human subjects is that subjects must give their *informed consent* to take part in the study. This means that they must understand the purpose of the study and any possible benefit or harm (possible harm arising, for example, because of the occurrence of adverse side effects of an otherwise beneficial drug) that they may experience as a result of taking part in the study. This being so, suppose that in our hypothetical example extremely obese patients decline to take part in the study. Suppose also that drug B is indeed more effective than drug A in controlling hypertension amongst non-obese or moderately obese patients, but not in severely obese patients. This phenomenon is illustrated in Figure 3.3, where different plotting symbols are used to indicate the hypothetical results that would have been obtained from patients who in fact declined to take part in the study; amongst these hypothetical results, there is no clear separation between the results for drugs A and B. Analysis of the data actually obtained would, as we have already seen, lead to the simple conclusion that drug B is superior to drug A. Had it been ethically acceptable for BMI to have been a design variable, the experiment could have covered the full range of BMI, and the more complex truth would have emerged: that drug B is only superior to drug A for non-obese or moderately obese patients.

Although the blood-pressure example is synthetic, we will see in later chapters that the phenomenon it illustrates is not unrealistic. Also, the hypothetical scenario described above is itself only a simple example of a more insidious but widespread phenomenon, *selection bias*. This can arise whenever some of the intended data are missing, for reasons related to the phenomenon under investigation. In our hypothetical example comparing two anti-hypertensive treatments, selection bias arose because participation in the study was voluntary and a subject's willingness to participate was related to their body mass index, which in turn was predictive of their blood pressure.

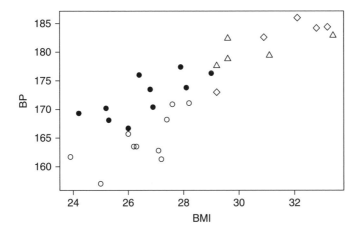

Fig. 3.3. Systolic blood pressure and body mass index for 30 subjects. Ten subjects were assigned drug A and agreed to participate in the study (solid dots), ten were assigned drug B and agreed to participate (open circles), five were assigned to drug A but declined to participate (triangles), five were assigned to drug B but declined to participate (diamonds).

3.2 Probability

Statisticians use probability theory as a precise measure of uncertainty. The intuitive idea of probability is a very familiar one. We deal with uncertainty intuitively in everyday decision-making. Is it likely to rain today? If so, I should wear a coat. Lancaster City have been drawn against Manchester United in the FA Cup. A win for Lancaster is unlikely.

To put this idea on a more formal footing, one approach is to imagine that something may or may not happen as a result of your doing something *that you can repeat indefinitely under identical conditions*. The usual, boring but easily understood, example is coin tossing. When I toss a coin it will or will not land 'heads up'. If I toss it four times, I could get 0, 1, 2, 3 or 4 heads and I might be mildly surprised if I got 0 or 4. If I toss it 10 times and get 9 or 10 heads, I begin to suspect that the coin is biased in favour of heads, but 7 out of 10, or a proportion 0.7, of heads would not be particularly surprising. However, if I toss it 1000 times and get a proportion 0.7 of heads, I am pretty confident that the coin (or the way I toss it) is biased. Figure 3.4 illustrates this. The proportion of heads in a sequence of tosses of a fair coin will fluctuate around the value 0.5, but the fluctuations will decrease in size as the number of tosses increases.

We define the *probability* of an event as the limiting proportion of times that the event occurs in an indefinitely long sequence of independent trials. This immediately begs the question of what, precisely, we mean by 'independent', although, as we shall see shortly, it means roughly what you

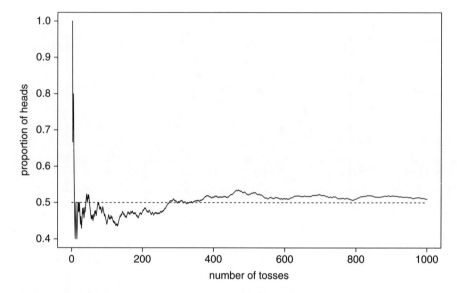

Fig. 3.4. The cumulative proportion of heads in a simulated sequence of 1000 tosses of a fair coin.

would expect it to mean. For this and other reasons a rigorously minded mathematician might not accept our 'definition' of probability, but for practical purposes it will serve us well enough.

Now let's get back to what we mean by 'independence'. We define two events as being *independent* if knowledge of whether one has occurred does not alter the probability of the other occurring. For example, in our hypothetical comparison of two anti-hypertensive drugs, the average of the 10 BP values in Table 3.1 for people given drug A is 172.1. Because the 10 values were obtained from 10 different people there is no reason to suppose that knowing that the first person has a BP value greater than 172.1 would be of any help to us in predicting whether the second person would also have a BP value greater than 172.1. The 10 subjects yield 10 *independent* pieces of information. Now observe that the first person's BMI is 27.9, which is above the average value, 26.6, of BMI for these same 10 people. Is it coincidental that a person's BP and BMI should *both* be above average (or, by the same token, both below average)? Probably not. Overweight people are more likely to suffer from hypertension: the two events 'above average BMI' and 'above average BP' are not independent.

A variation on our hypothetical example is the following. Suppose that each person provides not one but two values of BP, taken a day apart. Now we have 20 values of BP from people given drug A. But we do not have 20 independent values. A person who gives an above-average value on day 1 is likely also to give an above-average value on day 2.

Failure to recognize the inherent *dependence* between repeated measurements on the same experimental unit is one of the most common sources of fallacious statistical reasoning in the analysis of experimental data. Conversely, ensuring *independent* replication is a cornerstone of good experimental practice.

When the outcome of an experiment has only two possible values, as with a coin toss, its statistical properties are completely described by a single probability, say the probability of heads. If the probability of heads is denoted by p, then the probability of a tail is necessarily $1 - p$, because the outcome *must* either be heads or tails, and a probability of 1 corresponds to certainty. For example, if the coin is fair, then $p = 1 - p = 0.5$, whereas a coin biased in favour of heads has $p > 0.5$.

When the outcome has three or more possible discrete values, its statistical properties are described by a set of probabilities, one for each of the possible values, such that their sum is one. We call this set of probabilities, p_i, say, a *probability distribution*. Outcomes of this kind include counts and categorical variables. An example of a count would be the number of rainy days next week, in which case i can take values $0, 1, \ldots, 7$. Another would be the number of cancer-related deaths in the UK next year; strictly this must be a finite number, but its upper bound is hard to determine exactly. For a variable like this, the usual strategy for describing its statistical properties is to specify a probability distribution that allows arbitrarily large outcomes, but with vanishingly small probabilities.

The two requirements for any probability distribution are that the probabilities p_i are non-negative and that their sum is 1. When the outcome can take any value over a continuous range, as applies for example to blood pressure or body mass index, we encounter a modest paradox. Suppose, for example, that blood pressure is recorded to the nearest mm Hg, as in Table 3.1, then we might accept that the probability of the outcome $BP = 170$, might be of the order of 0.05 or so (the precise value is immaterial for our current purpose). A superior instrument might be capable of recording blood-pressure to an accuracy of 0.1 mm Hg, in which case logic suggests that the probability of the outcome $BP = 170.0$ should be of the order of 0.005. An even better instrument might ... and so on. Apparently, the probability of obtaining any specific outcome must be zero, yet the probability of *some* outcome must be 1. One way to resolve the paradox is to insist that there is a limit to the accuracy with which any outcome can be measured, and it can therefore be described by a discrete set of probabilities p_i. This is rather inelegant, because it prevents a straightforward answer to a simple question like: what is the distribution of blood-pressure values over the UK population?

A better solution is to describe the distribution of a continuously varying outcome by a smooth curve like the one in Figure 3.5, called a *probability density function*, usually abbreviated to pdf. A pdf must take

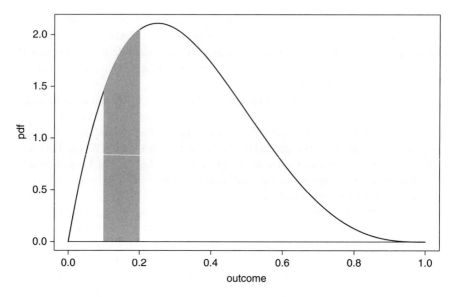

Fig. 3.5. A probability density function, representing an outcome that can take any value between zero and one. The area of the shaded region gives the probability that the outcome lies between 0.1 and 0.2.

only non-negative values, and the total area under the curve must be one. Then, the area under the curve between any two values, a and b, say, gives the probability that the outcome will take a value somewhere between a and b. For example, in Figure 3.5 the area of the shaded region is the probability that the outcome lies between 0.1 and 0.2. Notice that the value of a pdf can be bigger than one: probabilities are given by areas, *not* by the values of the pdf.

3.3 Statistical inference

In everyday language, *inference* and *deduction* are near-synonyms. In scientific language, there is a crucial difference. Inference is the process of drawing conclusions from evidence, whilst admitting the possibility that these conclusions may not be correct. Deduction is the process of logical argument from a premise to a necessary consequence. The term *statistical inference* refers to the process by which inferences are drawn from data. One key idea in statistical inference is that the degree of uncertainty associated with any conclusions drawn from a set of data can be expressed formally, and quantitatively, through the language of probability. Another, equally important but sometimes forgotten, is that the data can be regarded as a *sample* that is in some way representative of a larger *population* of scientific interest.

We will have much more to say about inference in later chapters, but to illustrate the essential ideas, we revisit our blood-pressure example. Specifically, consider the following ten numbers:

177, 169, 170, 167, 176, 174, 170, 174, 176, 168

Why should these numbers be of any interest to anybody? Well, they are values of blood pressure obtained from 10 people who have been given a drug (drug A) that claims to help control the blood pressure of hypertensive patients. So they are certainly of interest to the 10 people themselves. But if we believe that the results for these 10 people might be representative of what the drug would do for hypertensive patients in general, they might also be of interest to other patients and to their treating physicians. We call these 10 people a *sample* drawn from the *population* of all hypertensive patients who might be candidates for treatment with drug A.

The best way to ensure that a sample is representative is to draw it at random. The formal definition of a random sample is as follows. First, identify the population of interest, and label its members $1, 2, \ldots, N$. Then choose your required sample size n (in our example, $n = 10$) and pick n numbers at random from $1, 2, \ldots, N$. The members of the population whose labels have been picked in this way constitute a *random sample* from the population. This definition immediately runs into some difficulty if you can't explicitly label every member of the population of interest: how exactly would you identify all potential recipients of an anti-hypertensive drug? But the definition is still worth having, if only to encourage experimenters to ask themselves whether their method of acquiring their data is such that it can reasonably be assumed to behave *as if it were* a random sample.

The next idea we need to understand is that of estimation. We take samples because we have neither the time nor the money to examine whole populations. A numerical summary of the resulting sample, for example its average value or the proportion of values less than 170, is of interest if it provides a reasonable *point estimate* of the corresponding summary of the population. The average value of our ten blood-pressure values is $(177 + 169 + \cdots + 168)/10 = 172.1$. How do we know if this is a 'reasonable' estimate of the population average? Presumably, we would like to be confident that it is close to the population value, where what we mean by 'close' depends on the context. For blood pressure, an estimate with a likely error of up to 1 mm Hg either way would be accurate enough for any practical purpose, whereas one with a likely error of up to 20 mm Hg either way would be pretty useless. So, before deciding whether to trust our point estimate, we need to assess its likely error, which we do by converting it into an *interval estimate*. Here are another 10 hypothetical values of blood pressure sampled from a different population:

173, 174, 171, 173, 170, 172, 174, 170, 172, 172

These also have an average value of 172.1, but there is an obvious difference from the previous set of 10: they show much smaller variation about the average value. As a result, you would probably trust the estimate 172.1 more in the second case than in the first, and you would be right to do so. Using methods that we will discuss in more detail in later chapters, we can be 95% confident that the mean blood pressure in this second population lies somewhere in the range $(171.1, 173.1)$. In contrast, we can only be 58% confident of the same claim for the first population; to be 95% confident, we would have to widen the range to $(169.5, 174.7)$.

Now suppose that we want to assess whether the two drugs A and B differ in their effectiveness. Using the 20 values of blood pressure in Table 3.1, and again using methods that we will describe in detail in later chapters, we can be 95% confident that the difference between the population-wide average blood pressures achieved by the two drugs lies in the range $(3.4, 11.2)$ in favour of drug B, i.e., drug B produces the lower average, by somewhere between 3.4 and 11.2 mm Hg.

An interval estimate supplies a plausible range of values for a quantity of scientific interest. In our blood-pressure example, the interval estimate of the difference between the average values achieved by the two drugs suggests, amongst other things, that drug B is superior to drug A because the plausible range contains only positive values. There are at least two possible reactions to this.

One reaction is to question whether 95% confidence is good enough. What if we insist on 99% confidence? In that case, our interval estimate would widen to $(2.0, 12.6)$. How about 99.9% confident? Now, the interval becomes $(0.0, 14.5)$. So, if we accept a level of confidence up to 99.9%, we conclude that drug B is better than drug A, whereas if we insisted on higher levels of confidence than 99.9%, our interval estimate would include both positive and negative values, and the comparison would have to be declared inconclusive. We could also do the calculation the other way round, i.e., find the level of confidence so that one end of the interval estimate is exactly zero and report this 'critical' confidence level – the higher it is, the more inclined we are to believe that drug B really is better than drug A. Approaching the problem this way is called *testing* the hypothesis that there is no difference in the ability of the two drugs to achieve lower average blood pressure. Conventionally, if the critical confidence level is bigger than 95%, the test is said to have given a *statistically significant* result. Also conventionally, and confusingly, the result of the test is reported not as the critical confidence level, c say, but as $p = 1 - c/100$, the so-called *p-value*. So in our hypothetical example, $p = 1 - 99.9/100 = 0.001$.

In the authors' opinion, a more sensible reaction to the interval estimate $(3.4, 11.2)$ is to ask a doctor two questions. Would a difference of 11.2 mm Hg in average blood pressure be clinically useful? Would a difference of 3.4 mm Hg in average blood pressure be clinically useful? If the answer to both is 'no' then the drugs are equally good for practical purposes. If the answer to both is 'yes' then drug B is superior to drug A. One 'yes' and one 'no' means that the comparison is inconclusive, and we need more data. This still leaves us open to the criticism that using a 95% confidence interval is no more than a convention, but at least it focuses on the practically relevant question of how different the drugs really are, rather than on the somewhat academic question of whether or not they differ by what might be an arbitrarily small amount.

3.4 The likelihood function: a principled approach to statistical inference

Probably the most widely used way of reporting the results of a statistical analysis is the *p*-value. Most statisticians would prefer to see the emphasis placed on estimation, with results reported as confidence intervals rather than as *p*-values, a view that is beginning to be reflected in the editorial policies of some scientific journals; see, for example, Gardner and Altman (1986). In fact, there is a close operational link between confidence intervals and *p*-values. To explain this, we need to revisit and extend the concept of probability as described in Section 3.2.

Suppose that we wish to estimate the prevalence of a particular disease in a particular population, using a completely reliable diagnostic test. If ρ (the Greek letter rho) denotes the prevalence of the disease in the population, then ρ is also the probability that a randomly selected member of the population will test positive. Suppose now that we test 10 randomly selected members of the population and, writing $-$ and $+$ for a negative and positive result, respectively, obtain the sequence

$$\mathcal{D} = +--++-+++-$$

How should we estimate the prevalence? Setting aside, temporarily, the obvious answer, namely $6/10 = 0.6$ or, expressed as a percentage, 60%, we first derive an expression for the probabilities of each of the possible sequences $\mathcal{D}$ that could have been observed.

If, for the sake of argument, the prevalence is indeed 0.6, then there is a probability 0.6 that the first test result will be positive, a probability 0.4 that the second test result will be negative and therefore a probability $0.6 \times 0.4 = 0.24$ that the first two test results will be positive and negative, respectively. To understand why we have to multiply the two probabilities, imagine splitting a long sequence of test results into successive pairs. Then, we would expect 60% of the pairs to begin with a positive result, and of these 60%, 40% to be followed by a negative result, i.e., we expect 24% of the pairs to be $+-$. By the same argument, we expect

24% of the pairs to be −+, 16% to be −− and 36% to be ++. Note that these four percentages add to 100, as they must because they cover all of the possible outcomes. A crucial assumption in this argument is that the results of successive tests are *independent*, meaning that the probability of the second test giving a positive result is 60%, whether or not the first test gives a positive result. That seems a reasonable assumption in our hypothetical example, but it would not be so if, for example, the pairs of test results were from siblings and the disease was genetic in origin.

Of course, we don't know the true prevalence: in the terminology of Chapter 2, ρ is a *parameter*. But for any hypothetical value of ρ, we can repeat the above argument to obtain the following table of probabilities:

Outcome	Probability
−−	$(1-\rho) \times (1-\rho)$
−+	$(1-\rho) \times \rho$
+−	$\rho \times (1-\rho)$
++	$\rho \times \rho$

Also, the argument extends in the obvious way to a sequence of more than two test results. In particular, the probability associated with *any* sequence $\mathcal{D}$ that contains four negatives and six positives is

$$P(\mathcal{D}; \rho) = (1-\rho)^4 \rho^6. \tag{3.1}$$

The notation $P(\mathcal{D}; \rho)$ looks a bit cumbersome, but emphasizes that the expression on the right-hand side is derived from a particular set of data, $\mathcal{D}$, but is also a function of a mathematical variable, ρ. The left-hand panel of Figure 3.6 plots this function. Not coincidentally, it attains its maximum value when $\rho = 0.6$. We call the function $P(\mathcal{D}; \rho)$ the *likelihood function* for ρ given the data $\mathcal{D}$, often abbreviated to the likelihood for ρ and written $\ell(\rho)$. The right-hand panel of Figure 3.6 plots the corresponding *log-likelihood* function, $L(\rho) = \log \ell(\rho)$. The same value of ρ necessarily maximizes both $\ell(\rho)$ and $L(\rho)$, and is called the *maximum likelihood estimate* of ρ, usually written $\hat{\rho} = 0.6$ to emphasize that it is an estimate, and not the true value of ρ.

At this point, we forgive any reader who asks: why go through all this rigmarole to answer such a simple question? But bear with us. The great strength of likelihood-based methods is that they can be used to give numerical solutions to more complicated problems for which there is no intuitively obvious way to estimate model parameters. More than this, they also give a principled approach to constructing confidence intervals, and to testing hypotheses. Let $L(\theta)$ denote the log-likelihood function for a single parameter θ in a statistical model, and $\hat{\theta}$ the maximum likelihood estimate

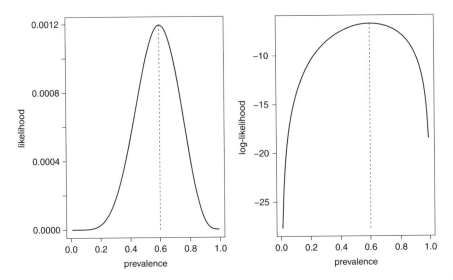

Fig. 3.6. The likelihood function (left-hand panel) and log-likelihood function (right-hand panel) for the unknown prevalence, ρ, given a sequence of ten tests with six positive and four negative results.

of θ. Then, for a very wide class of statistical models, including all of those used in this book, the set of values of θ for which $L(\theta) > L(\hat{\theta}) - 1.92$ is an approximate 95% confidence interval for θ, and a hypothesized value $\theta = \theta_0$ is rejected at the 5% significance level (i.e., p-value less than 0.05) if it is *not* inside this confidence interval, i.e., if $L(\theta_0) < L(\hat{\theta}) - 1.92$. Figure 3.7 illustrates graphically how this likelihood-based confidence interval is calculated, using the same data and model as in Figure 3.6.

For models with two or more parameters, the log-likelihood is a multidimensional surface. This is difficult to visualize, but the same method applies except that the number 1.92 is replaced by a larger number whose value depends on the number of parameters in the model; for two and three parameters, the relevant values are 3.00 and 3.91. This approach to hypothesis testing is called *likelihood ratio* testing. Many widely used statistical tests are special cases of a likelihood ratio test. Conventionally, the likelihood ratio test statistic is defined as $D = 2\{L(\hat{\theta}) - L(\theta_0)\}$ where θ now denotes the complete set of parameters, θ_0 is a constrained version of θ that fixes the numerical values of m parameters, and the hypothesis $\theta = \theta_0$ is rejected if $D > c_m(0.05)$ where $c_m(0.05)$ is called the *5% critical value*, and 0.05, or 5%, is called the *prescribed significance level* of the test. Table 3.2 gives values of $c_m(p)$ for $m = 1, 2, 3, 4, 5$ and $p = 0.05, 0.01, 0.001$.

This method of testing hypotheses is very general and widely used. For most realistically complex statistical models it is impossible to obtain explicit expressions for the maximum likelihood estimates. Instead, the

Table 3.2. Critical values of the likelihood ratio test statistic D for number of parameters m and prescribed significance level p.

p	\multicolumn{5}{c}{m}				
	1	2	3	4	5
0.05	3.84	5.99	7.81	9.49	11.07
0.01	6.63	9.21	11.34	13.28	15.09
0.001	10.83	13.82	16.27	18.47	20.52

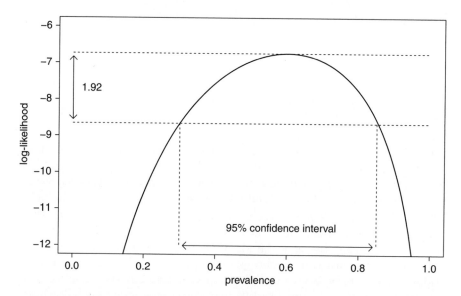

Fig. 3.7. The log-likelihood function for the unknown prevalence, ρ, given a sequence of ten tests with six positive and four negative results, and construction of the likelihood-based approximate 95% confidence interval for ρ.

estimates are found numerically by maximizing the log-likelihood function. The maximized value of the log-likelihood, $L(\hat{\theta})$, is an automatic by-product, and the numerical value of the likelihood ratio statistic D can therefore be found by maximizing the log-likelihood with and without the constraints imposed by the hypothesis $\theta = \theta_0$.

Anybody who wants to use statistical methods to analyse their data is confronted with a plethora of seemingly unrelated techniques for calculating confidence intervals or p-values in particular situations. This makes statistical method look a bit like cookery, with a recipe for every occasion.

But most of the better recipes are particular examples of the likelihood method. They were developed at a time when computers were unavailable and explicit formulae were needed to enable hand calculation of the required results. Now that computers are ubiquitous, and fast, restricting the statistical recipe book to explicit formulae is unnecessary. Far better, in the authors' opinion, to adopt a principled approach to statistical analysis, in which the old 'which test can I use on my data?' mantra is replaced by 'how can I use statistical thinking to help me get the best possible answer to my scientific question?' The likelihood function plays a crucial role in the new mantra, because it provides a principled, and almost universally applicable, approach to inference.

3.5 Further reading

The operational definition of probability given in Section 3.2 is called the *relative frequency* definition. Mathematicians favour an *axiomatic* definition, by which they mean that they assert a minimal set of properties that something called probability should have, and deduce as many consequences of these assertions as they can. It turns out that the minimal set really is very minimal, essentially no more than requiring that probabilities should be numbers between zero and one, together with a mathematical formalization of the everyday meaning of probability, to the effect that relaxing a target cannot make it harder to hit. I don't know what the probability is that, sometime in your life, you will run 1 kilometre in less than 3 minutes, but it cannot be smaller than the probability that you will ever run 1 kilometre in less than 2 minutes 30 seconds, because you can achieve the former *either* by achieving the latter *or* by running 1 kilometre in a time between 2 minutes 30 seconds and 3 minutes. The axiomatic approach to probability was developed by the Russian mathematician Andrey Kolmogorov (1903–1987). Relatively easy introductions to probability theory, albeit aimed at first-year undergraduate mathematics students rather than general science students, include Chapter 5 of Chetwynd and Diggle (1995) or, in considerably more detail, McColl (1995).

A philosophically very different view of probability and its role in scientific inference is the idea that probability measures a person's degree of belief in a proposition, for example that drug B is superior to drug A in reducing average blood pressure. Under this view, there is no objection to different people having different beliefs and therefore assigning different probabilities to the same thing. However, to convert what would otherwise be superstition into scientific method, this view of probability theory requires its users to respond rationally to evidence. So, to continue the example, it is acceptable for one of us to believe strongly (say, probability 0.9) that drug A is superior to drug B and for the other of us to be more sceptical (say, probability 0.5), *before we have seen any data*. But faced with the data in Table 3.1 we would both be forced to revise our probabilities downwards in line with the evidence in favour of drug B that the data provide. And we would both do so by working to the same set of well-defined mathematical rules. This mode of statistical inference is known as *Bayesian* inference, in honour of the Reverend Thomas Bayes (1702–1761). Bayes was not an advocate of Bayesian inference, which only came into being long after his death, but he set out the fundamental mathematical result concerning revision of a probability in the light of empirical evidence that became the foundation on which Bayesian inference was built. Lindley (2006) is a thought-provoking, if challenging, account of these ideas.

4
Exploratory data analysis: gene expression microarrays

4.1 Gene expression microarrays

Recent technological developments in biology have led to the generation of enormous volumes of data. An example is the technology of gene expression microarrays. Figure 4.1 shows an example of a cDNA, or 'red-green' microarray, so-called because the colour of each spot on the array estimates the ratio of expression levels for a particular gene in response to each of two stimuli, captured by the intensity of red and green colour, respectively. A cDNA array is a glass slide on which tiny amounts of genetic material have been deposited by a robotic device at each of several thousand positions. More modern microarray technologies achieve greater numbers still.

In very simple terms, measuring gene expression is interesting because individuals within a species show genetic variation in the extent to which individual genes or groups of genes within the species' genome are *expressed*, or 'switched on'. Until relatively recently, if a scientist wanted to understand which genes were switched on by a particular stimulus, such as exposure to a disease-causing organism, they needed to select individual candidate genes for detailed study. Using microarrays, many genes can be investigated in parallel. The name embraces a number of different technologies. Our example uses a technology developed by the Affymetrix company, and concerns an investigation into the phenomenon of calcium tolerance in the grass, *Arabadopsis thaliana*. The data were kindly provided to us by Dr Bev Abram (Biological Sciences, Lancaster University).

Dr Abram's experiment was of a type known as a 2×2 factorial. Four different sets of experimental conditions, or *treatments* were used, consisting of all four combinations of two *factors*. The first factor was the strain of *Arabadopsis* used, one bred to be calcium resistant, the other a wild-type strain. The second factor was the calcium challenge, low or high, to which the plants were exposed. Each of the resulting four treatments was replicated three times, giving a total of 12 experimental units.

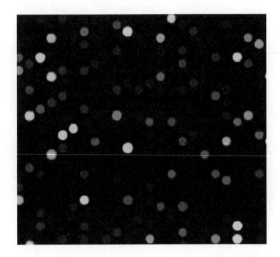

Fig. 4.1. A microarray image. Each spot on the array represents a gene. The colour of the spot codes for the ratio of expression levels of that gene in response to two different stimuli, from bright green at one extreme to bright red at the other. This figure is reproduced in colour in the colour plate section.

DNA was extracted from plant material taken from each experimental unit, and processed using an Affymetrix microarray. For each of the 12 samples of plant material, the microarray technology measures the expression levels for each of 22,810 genes. The dataset therefore consists of $22{,}810 \times 12 = 273{,}720$ numbers. The aim of the experiment was to identify for more detailed investigation a smaller number of genes that showed a response to the calcium challenge in the calcium-resistant strain but not in the wild-type strain.

Although we have described the data as 'measured' expression levels, it would be more accurate to call them 'estimated' levels. Impressive though it is, the microarray technology is subject to many sources of imprecision in its determination of levels of genetic activity. Some of these are an inescapable consequence of natural biological variation between different plants, whilst others stem from technical variation in the performance of the array technology. For this reason, microarray data are usually pre-processed in various ways, in an attempt to remove technical artifacts, before they are analysed statistically. The data that we use in this chapter are of this kind. Even so, they contain both biological and technical variation in an unknown mix, and it is this which makes their interpretation challenging.

Table 4.1 shows the data from four randomly selected genes. Some things are immediately obvious: Gene 3 shows much higher expression levels than Genes 1, 2 or 4, over all four treatments and all three replicates; there

Table 4.1. Pre-processed gene expression data for four randomly selected genes in the *Arabadopsis* experiment: $1 = 253023.at$, $2 = 244928.s.at$, $3 = 262200.at$, $4 = 250740.at$.

Treatment		Gene			
Strain	Ca challenge	1	2	3	4
Ca resistant	low	5.208	6.469	12.924	7.854
		6.649	6.575	15.234	10.226
		7.122	7.080	18.027	9.111
	high	5.031	7.714	16.699	8.308
		4.755	8.429	15.186	8.688
		6.053	6.998	15.709	12.249
wild type	low	6.410	6.357	11.772	8.741
		5.482	6.278	16.407	8.395
		5.180	5.906	24.235	8.913
	high	5.211	6.874	17.471	8.645
		6.710	7.870	14.657	7.255
		5.564	3.872	20.247	13.179

is considerable variation between the results for any one gene and/or any one treatment. Other aspects of the data are more subtle. In particular, the biological interest in this experiment lies in identifying genes which appear to show differential expression levels between the different treatments. It is difficult to assess these patterns by direct inspection of the data in Table 4.1, and it would clearly be impossible to do so for all 22,810 genes in the complete dataset.

Although these data can be explored using simple and familiar graphical methods, their volume demands careful choices to be made in the way the results are displayed. Indeed, an underlying theme throughout this chapter is that whilst modern computer graphics facilities make it easy to produce graphs of data, the default graphs produced by statistical software packages are often unsatisfactory.

In this and many other examples, the immediate goals of exploratory data analysis are: to describe the typical patterns of variation in the data; and to highlight exceptions to the typical patterns. An indirect goal is often to suggest what kind of statistical model would be appropriate for making formal inferences, as was the case in our discussion of the undergraduate physics lab experiment in Chapter 2. However, this need not always be the case. Sometimes, an exploratory analysis will be followed by the design of an experiment to generate an entirely new set of data for analysis, or by a confirmatory investigation of a non-statistical kind. In the following sections, we shall use subsets of the complete data to illustrate different ways of exploring a dataset. From a scientific perspective, there is of course little or no value in analysing arbitrary subsets of the data. Our reason for doing so here is to emphasize that different kinds of graphical presentation are suited to exploring small or large datasets. To a lesser extent, the same applies to tabular methods of presentation, and in particular to the need for summarization, rather than direct inspection of the data themselves.

4.2 Displaying single batches of data

The simplest structure for a dataset is a *single batch*. This consists of a set of data, say $y_1, y_2, \ldots, y_n$, with no internal structure of any kind. Amongst other things, this implies that there is no significance to the order in which the individual values y_i are labelled, nor any rational basis for dividing the data into subgroups prior to statistical analysis. For illustration, we shall consider the values obtained from the first of the 12 arrays in our gene expression data. In doing so, we are deliberately ignoring any biological information that may or may not be available to define subgroups of genes, for example, according to their biological function.

For small batches, a primitive but useful display is the *dot-plot*. To illustrate, Figure 4.2 shows a dot-plot of expression levels recorded on the first array for a random sample of 25 genes from the 22,810 available. As the name implies, each datum is shown as a dot, which is plotted at the appropriate point of an axis corresponding to the measured gene expression level.

The plot shows clearly that most genes exhibit relatively low expression levels, with a long upper 'tail' composed of a few genes with very high expression levels. Distributions of this kind are called *positively skewed* distributions (negatively skewed distributions have a long lower tail), and can be difficult to interpret because most of the variation in the data is condensed into the clump of dots at the left-hand end of the plot. It often helps to plot positively skewed distributions on a logarithmic scale. In the context of the gene expression data, it is also the case that biologists prefer

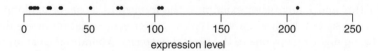

Fig. 4.2. Dot-plot of measured gene expression levels for a random sample of 25 *Arabadopsis* genes.

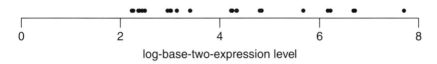

Fig. 4.3. Dot-plot of log-base-two-transformed gene expression levels, for a random sample of 25 *Arabadopsis* genes.

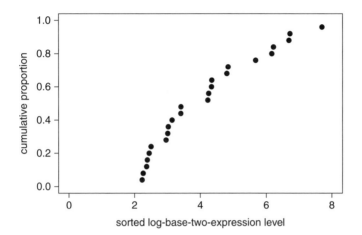

Fig. 4.4. Cumulative plot of log-base-two-transformed gene expression levels, for a random sample of 25 *Arabadopsis* genes.

to work with logged expression values, in particular using log-base-two, giving a scale on which each unit change corresponds to a two-fold change in expression. Figure 4.3 shows the same data as did Figure 4.2, but now in terms of the log-transformed data. The distribution is still positively skewed, but less so than before. The individual values are also easier to see.

Another way to display a small batch of data is to use a *cumulative plot*. If we order the values, say $y_1, y_2, \ldots, y_n$, from smallest to largest, then a cumulative plot is a plot of the cumulative proportions, i/n, against the ordered values y_i. Figure 4.4 shows a cumulative plot of the same data as in Figure 4.3. The positively skewed shape of the distribution now shows up as a pronounced curvature in the plot.

Most people find cumulative plots difficult to interpret on first acquaintance. But it is worth getting to know them better, as they can be more effective than dot-plots in revealing distributional shape in small batches of data.

Dot-plots become cluttered for larger samples of data. Cumulative plots can still be used to avoid the clutter, but a non-cumulative solution is also available. Instead of plotting individual values, we collect these into

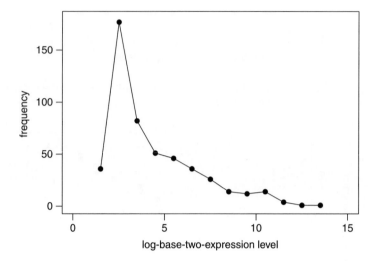

Fig. 4.5. Frequency polygon of log-base-two-transformed gene expression levels, for a random sample of 500 *Arabadopsis* genes.

ranges, called *bins* and use the vertical axis of the plot to represent the numbers in each bin. Figure 4.5 shows an example for a random sample of 500 *Arabadopsis* genes, using bins of unit width; for example, the first plotted point, at $x = 1.5$, $y = 37$, indicates that 37 of the genes had measured log-expression levels in the range one to two units. Connecting the dots by lines is done purely for visual effect, but explains why a plot of this kind is sometimes called a *frequency polygon*. The plot very clearly shows the positively skewed nature of the underlying distribution, with a sharp peak of log-expression levels in the range two to three units.

Two common variants of the frequency polygon are the *bar-chart* and the *histogram*. Figure 4.6 shows these two plots for the same data as in Figure 4.5. The only difference between the two is that in the bar-chart the vertical bars are separated, whereas in the histogram they are contiguous. This may seem a small point, but its purpose is to signal that in a bar-chart, the plotted frequencies should correspond to strictly discrete values in the data, whereas in a histogram the plotted frequencies correspond to data values which in fact vary on a continuous scale but have been grouped into bins for convenience. In other words: use a bar-chart or frequency polygon when the data values are discrete; use a histogram when, as is the case here, they are continuous.

Figure 4.7 shows a histogram of all 22,810 log-transformed expression levels from the first array in Dr Abram's experiment. Notice how, with this very large sample size, the width of each bin has been reduced to 0.5 and the plotted distribution is very smooth. It shows a single peak in

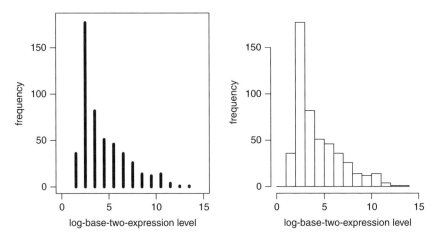

Fig. 4.6. Bar-chart (left-hand panel) and histogram (right-hand panel) of log-base-two-transformed gene expression levels, for a random sample of 500 *Arabadopsis* genes.

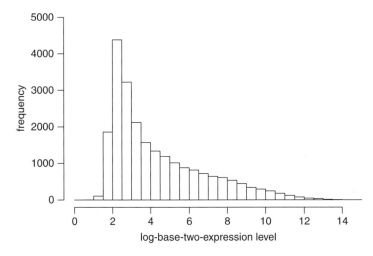

Fig. 4.7. Histogram of log-base-two-transformed gene expression levels, for 22,810 *Arabadopsis* genes.

the bin corresponding to values between 2.0 and 2.5 and no secondary peaks. This illustrates how batches with as many as several hundred data values can still show features that are a by-product of random sampling variation rather than reflecting genuine features of the underlying population.

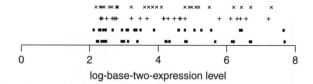

Fig. 4.8. Superimposed dot-plots of log-expression levels of 25 genes, in each of the four treatment groups: Ca-resistant strain under low Ca challenge (solid dots); Ca-resistant strain under high Ca challenge (open circles); natural strain under low Ca challenge (+), natural strain under high Ca challenge (×).

4.3 Comparing multiple batches of data

Multiple batches of data arise naturally when experiments are performed to compare results under two or more different treatments. The graphical methods described above for single batches can be adapted to multiple batches in at least two different ways: either by arranging graphical displays of each batch in a multi-panel plot or by superimposition on a single plot, using plotting symbols, line styles or colours to distinguish batches.

Whichever method is preferred, it is worth thinking carefully about the overall design of the graphical display. For multiple panels the choice of separate or common ranges for the x and y axes of each panel, and their layout on the page, can help or hinder interpretation. Superimposed graphics can easily become cluttered and difficult to read. We now use versions of Bev Abram's microarray data to illustrate these choices, in each case equating batches to the four different treatments.

For small samples, a set of superimposed dot-plots is a simple, and often effective choice. Figure 4.8 shows superimposed dot-plots of the measured log-expression levels for a sample of the same 25 genes in each of the four treatment groups, distinguished by different plotting symbols. There are no obvious differences amongst the four distributions.

Figure 4.9 shows the same data as a set of superimposed cumulative plots, this time with successive points connected by line segments to minimize clutter. The clear conclusion is that the four distributions are almost identical. This is to be expected. Few genes are likely to be involved in regulating the plants' response to calcium exposure, hence the likelihood that a random sample of 25 out of 22,810 would show differences in gene expression is remote.

There would be no obvious advantage to replacing Figure 4.9 by a multi-panel plot. For other forms of display, the choice is less clear-cut.

Figure 4.10 shows histograms of log-expression levels from each of the four treatments for a much larger sample of 500 genes. The two-by-two layout acknowledges the factorial structure of the four treatments, whilst the use of common scalings for the x and y axes makes it easy to see the

COMPARING MULTIPLE BATCHES OF DATA   41

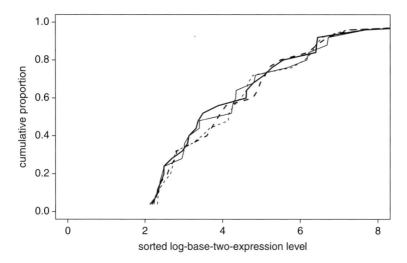

Fig. 4.9. Superimposed cumulative plots of log-expression levels of 25 genes, in each of the four treatment groups: Ca-resistant strain under low Ca challenge (thin solid lines); Ca-resistant strain under high Ca challenge (thick solid lines); natural strain under low Ca challenge (thin dashed lines), natural strain under high Ca challenge (thick dashed lines).

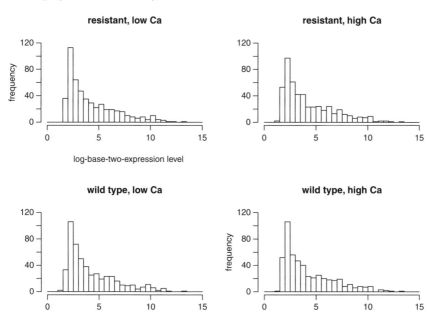

Fig. 4.10. Histograms of log-base-two-expression levels of 500 randomly selected *Arabadopsis* genes in each of four treatment groups. The two-by-two layout reflects the two-by-two factorial structure of the four treatments.

small differences in shape amongst the four histograms. A superposition of these same four historgams on a single plot (not shown) is very cluttered, making it difficult to see whether or not there are any interesting differences amongst the four treatments.

4.4 Displaying relationships between variables

A central concern of science is to understand *relationships* amongst two or more measured variables in an experiment. The simplest situation of this kind is when each run of an experiment generates a pair of measured values, say x and y. Replication of this experiment would result in a dataset consisting of n pairs, say $(x_i, y_i) : i = 1, \ldots, n$. For example, Table 4.2 shows measured expression levels of two genes in the *Arabadopsis* experiment.

A *scatterplot* of data pairs $(x_i, y_i) : i = 1, \ldots, n$ is simply a plot of the data with each pair of values drawn as a point in two-dimensional (x, y)-space. Figure 4.11 shows this for the data in Table 4.2. Notice that gene A

Table 4.2. Measured gene expression levels from two genes, A and B, in an experiment with 12 replicates.

Gene	Replicate											
	1	2	3	4	5	6	7	8	9	10	11	12
A	12.9	15.2	18.0	16.7	15.2	15.7	11.8	16.4	24.2	17.5	14.7	20.2
B	7.9	10.2	9.1	8.3	8.7	12.2	8.7	8.4	8.9	8.7	7.3	13.2

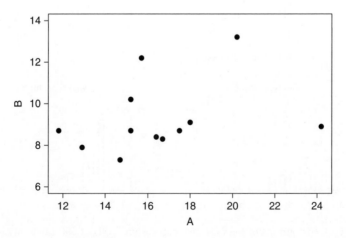

Fig. 4.11. Scatterplot of expression levels for two *Arabadopsis* genes in an experiment with 12 replicates.

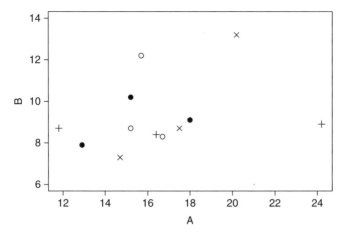

Fig. 4.12. Scatterplot of expression levels for two *Arabadopsis* genes in four experiments (distinguished by different plotting symbols) with three replicates each.

shows more variability in its expression levels over the 12 replicates than does gene B. Other than that, the plot is not very revealing – there is perhaps a hint of a positive association between the two genes, in the sense that when one is more highly expressed than average so is the other, and vice versa. If this were verified in a larger experiment, it would suggest a potentially interesting pattern of co-expression of the two genes.

The more astute reader may already have noticed that the data in Table 4.2 are simply the values, rounded to one decimal place, of the data in the final two columns of Table 4.1. So the data do not represent 12 replicates at all, but rather three replicates of four experiments conducted under different conditions. A more honest plot of these data is therefore Figure 4.12 where we have used different plotting symbols to differentiate the four different experimental conditions. Now, and with the important caveat that no definitive conclusions can be expected from such a small dataset, we see clear differences in the relationship between the two genes across the four experiments: in one case (+) the expression levels of gene A vary substantially over the three replicates whilst the expression levels of gene B are approximately constant; in another (o) the converse holds; in a third (•) there is a hint of weak positive association; in the fourth (×), the association appears stronger.

The lesson of this example is that the value of a single scatterplot of multiple batches of data can be enhanced by using different plotting symbols to differentiate between batches. Unsurprisingly, scatterplots are also much more useful for datasets that are too large to permit direct inspection of the data in tabular form. Figure 4.13 shows a different aspect of the *Arabadopsis* data. Now, each point on the plot represents

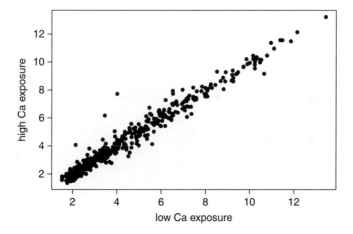

Fig. 4.13. Scatterplot of average log-based-two-transformed expression levels for 500 *Arabadopsis* in plants exposed to low or high levels of calcium.

the expression levels, averaged over three replicates, of a single gene under two different experimental conditions, exposure to low and high levels of calcium. The immediate message from the plot is that genes which are highly expressed under one experimental treatment tend also to be highly expressed under the other, and vice versa. Note also the high concentration of points towards the lower-left-hand corner of the plot. Most of the 500 genes show low levels of expression under both levels of calcium exposure. This is what we would expect, because we know that most of the genome consists of inactive genes.

Large datasets convey potentially rich messages, but their associated plots can become cluttered. A simple way to reduce clutter in a scatterplot that shows a strong overall association between the two variables is to *level* the plot. Instead of plotting the points (x_i, y_i) themselves, we plot their difference, $y_i - x_i$ against their average, $(y_i + x_i)/2$. Figure 4.14 shows the result of levelling Figure 4.13. The overall message of the plot is as before, but the effective magnification of the plot makes it easier to see some of the detail. In particular, recall that the experiment was designed to investigate whether some genes, and if so which ones, show enhanced expression when exposed to high, rather than low, levels of calcium, since these are likely to be the genes involved in the plant's internal calcium regulation system. Specifically, and remembering that expression levels are measured on a log-base-two-transformed scale, the 11 points that lie above the value 1 on the vertical axis of Figure 4.14 correspond to genes whose estimated expression levels have at least doubled in response to the high level of calcium exposure; these 11 genes are therefore candidates for further investigation of their possible involvement in calcium regulation.

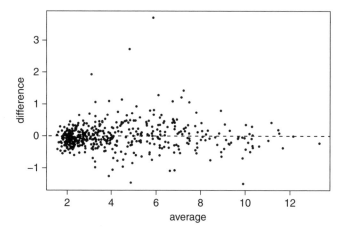

Fig. 4.14. Levelled scatterplot of average log-base-two-transformed expression levels for 500 *Arabadopsis* in plants exposed to low or high levels of calcium.

4.5 Customized plots for special data types

In the earlier sections we have described the more widely used ways of presenting data graphically. Some kinds of data need their own methods of graphical presentation.

4.5.1 *Time series*

A time series is a sequence of values $y_1, y_2, ..., y_n$ in which y_t, the tth member of the sequence, is the value of a measurement made at time t. The standard form of graphical presentation for a time series is a time-plot, which is simply a scatterplot of the points (t, y_t) with successive points connected by straight lines. Figure 4.15 gives an example in which y_t is the maximum temperature recorded at the Hazelrigg field station, near the Lancaster University campus, on day t running from 1 September 1995 to 31 August 1996. The strong seasonal effect is clear, but the plot also shows the pattern of short-term fluctuations in temperature around the seasonal trend; for example, the period between days 128 to 138 (6 to 16 January 1966) was unseasonally mild, with temperatures close to or above 10 degrees.

Figure 4.16 shows a much shorter time series plot, in this case of only the following six values:

$$87 \quad 73 \quad 69 \quad 54 \quad 69 \quad 52$$

Note that the plotted points are not equally spaced along the time axis. The data are from a study of the effectiveness of a drug treatment for schizophrenia. A measure of the severity of the subject's symptoms, PANSS

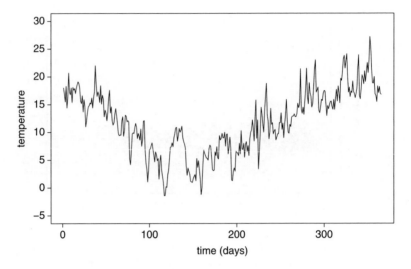

Fig. 4.15. Time series plot of maximum daily temperatures in degrees Celsius, at the Hazelrigg field station, near the Lancaster University campus, between 1 September 1995 and 31 August 1996.

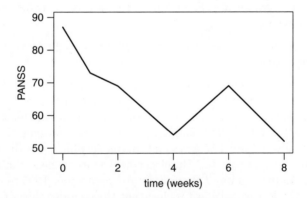

Fig. 4.16. PANSS measurements from a schizophrenia patient at times $t=$ 0, 1, 2, 4, 6 and 8 weeks since recruitment.

(Positive and Negative Symptom Score), was made at times 0, 1, 2, 4, 6 and 8 weeks following the subject's recruitment to the study.

The treatment seems to have improved this person's symptoms. But would we be confident in predicting an equally good result for other schizophrenia sufferers? Clearly not: we need to replicate the study. Figure 4.17 shows the data from 50 subjects, all of whom were treated with the same drug. Individual PANSS response profiles differ markedly between subjects; some show marked improvement over the eight-week follow-up

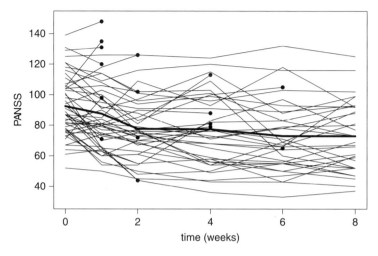

Fig. 4.17. PANSS measurements from 50 schizophrenia patients at times $t = 0, 1, 2, 4, 6$ and 8 weeks since recruitment. The thick line shows, at each follow-up time, the average PANSS score amongst patients who have not yet dropped out of the study.

period, others do not. The profile of average PANSS scores over time does show a steady decrease. A preliminary conclusion might be that the novel medication is beneficial on average, but not effective in all patients; the reduction in the mean PANSS score over the eight weeks, from 92.5 to 73.0, would certainly be regarded as a clinically worthwhile improvement. Note, however, some shortened series, highlighted by a solid dot at the end of each. These identify patients who withdrew from the study before eight weeks. The preponderance of above-average termination points suggests that the withdrawals tended to be the less well, in which case a crude comparison of initial and final mean PANSS scores could paint the novel medication in an unduly rosy light.

A plot like Figure 4.17 is called a *spaghetti plot*; the name suggests, often with good reason, a tangled mess. However, with a little care and the addition of a line showing the average response at each time point, the diagram gives us a reasonably clear picture. Data of this kind, consisting of relatively many short time series, are called *longitudinal data*. Early withdrawals are called *dropouts*. Drawing valid conclusions from longitudinal studies with dropout requires care if, as seems to be the case in this example, the dropout process is related to the phenomenon under investigation. General discussions of statistical methods for the analysis of longitudinal data include Diggle, Heagerty, Liang and Zeger (2002) and Fitzmaurice, Laird and Ware (2004).

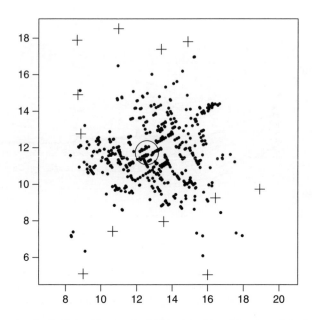

Fig. 4.18. London's Soho district in 1854 with the addresses of cholera patients marked as black dots. The location of the Broad Street pump is at the centre of the open circle. The locations of other pumps are marked as crosses.

4.5.2 *Spatial data*

Some kinds of data have an important spatial context, and are best presented as maps. Figure 4.18 is a famous early example. It shows the residential locations of cholera fatalities in London's Soho district at the time of the cholera epidemic of 1854, together with the locations of public water pumps. The concentration of cases in the centre of the map, on and around Broad Street, led Dr John Snow to conclude, correctly, that contaminated water from the Broad Street pump was the cause of the epidemic; the full story is told in Hempel (2006).

A dataset of the kind shown in Figure 4.18 is called a *spatial point pattern*; the information it gives is purely locational. Data in which each location carries with it a measurement is called a *marked* point pattern if both the locations and the measurements are of scientific interest, or a *geostatistical* dataset if only the measurements are of interest. Figures 4.19 and 4.20 show an example of each. The first is an example of a marked spatial point pattern. It is a map of the locations and sizes of trees in a portion of a mature forest in Saxony, Germany. An ecologist might be interested in understanding why the trees are where they are, and how their locations relative to each other affect their growth. The second is an example of a geostatistical dataset. It is a map of lead concentrations

CUSTOMIZED PLOTS FOR SPECIAL DATA TYPES 49

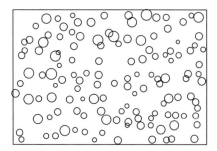

Fig. 4.19. Data on the locations and sizes of 134 spruce trees in a mature stand within a forested area of Saxony, Germany. The dimensions of the rectangular area are 56 by 38 metres. Circle radii are proportional to tree size.

measured from samples of moss gathered at the locations indicated in the province of Galicia, Spain. Both datasets have the same numerical format, namely a set of triples defining a measured value at a geographical location. The distinction between the two is that in Figure 4.20 the locations were chosen not by nature but by the scientists conducting the study, whose purpose was to estimate the overall pollution surface throughout the geographical region in question.

4.5.3 *Proportions*

Until now, all of our examples have considered data that are quantitative in nature: either a discrete count, or a continuous measurement. When the outcome of an experiment or observation is a qualitative piece of information, the resulting data are called *categorical*, or sometimes *unordered categorical* to emphasize their qualitative nature. A simple genetic example would be one in which a single locus carries a gene of type A or B, and a child therefore inherits from their parents one of three genotypes, AA, AB or BB. The standard form of graphical presentation of such data is as a *pie chart*. Figure 4.21 gives an example relating to a study of genetic variation in isolates of the bacterium *Campylobacter jejuni*, a common cause of gastroenteric infection in developed countries. *C. jejuni* has identifiably different genetic profiles in different host species. Wilson *et al.* (2008) compared the genetic profiles of isolates from human cases of campylobacteriosis with those of isolates from various animal hosts, and used these data to ascribe to each human case the most probable species-of-origin. Figure 4.21 shows the resulting distribution over the three most common species-of-origin, cattle, chicken and sheep, and the small proportion ascribed to a variety of all other species-of-origin. At the time the work was done, contaminated chicken was widely recognized as a major source of human cases, whereas the importance of cattle and sheep as host species was somewhat surprising.

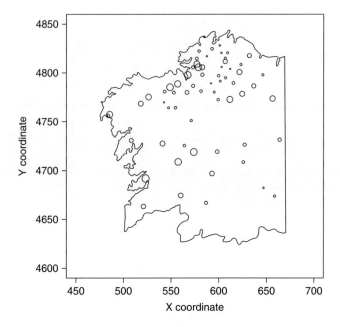

Fig. 4.20. Lead concentrations measured from samples of moss gathered at the 63 locations indicated in the province of Galicia, Spain. Circle radii are proportional to lead concentration.

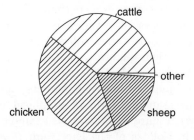

Fig. 4.21. Pie chart showing the proportions of human cases of campylobacteriosis ascribed by Wilson *et al.* (2008) to having originated in chicken, cattle, sheep or other species.

Pie charts are much beloved by management consultants. Statisticians use them sparingly, and only to describe categorical data.

4.6 Graphical design

Well-constructed graphical displays of data can convey a great deal of information clearly and succinctly. Default graphics produced by software

are often not well constructed. Common faults include clutter (a recurring word throughout this chapter) caused by thoughtless choice of plotting symbols or line styles, and unnecessary decoration that distracts from the important information in the diagram. Colour can be effective but tends to be overused; and graphic designers should remember that red–green is the most common form of colour blindness. A classic work on the design of statistical graphs is Tufte (1983). More technical discussions include Cleveland (1984) or Wainer (1997, 2005). Modern computing environments allow for easy animation, which has an obvious relevance to the display of data on time-varying phenomena and can also be used to good effect in visualizing high-dimensional data through the use of dynamic graphical displays. Cleveland and McGill (1988) is an edited collection of ideas in this area.

4.7 Numerical summaries

4.7.1 *Summation notation*

Many statistical operations on a set of data involve repeated addition of data values, or of quantities derived from data values. Summation notation is a very convenient and compact notation for defining these operations. Suppose, for example, a dataset consisting of the first two rows of Table 2.1, which we here reproduce as

i	t (sec)	d (cm)
1	0.241	10
2	0.358	40
3	0.460	70
4	0.249	10
5	0.395	45
6	0.485	75

Each row now contains two pieces of information, t and d, and an *index* variable, i, running from 1 to 6, the number of rows in the table. Suppose we want to calculate the sum of the six values of t. In summation notation, this is written as

$$\sum_{i=1}^{6} t_i,$$

in which $\sum$ is the *summation sign*, the subscript $i = 1$ and superscript 6 indicate the *range* of the summation and t_i denotes the value of t corresponding to index i; hence, for example, $t_2 = 0.358$. Summation notation is particularly useful for defining unambiguously more complex operations on data. For example, the slightly cumbersome instruction 'take the product of each pair of values t and d, add them up and divide by the sample size' reduces to $(\sum_{i=1}^{6} t_i d_i)/6$.

4.7.2 Summarizing single and multiple batches of data

The simplest and most widely used summary of a single batch of data is its mean, or average value. The *sample mean* of a batch of values $x_1, x_2, x_3, \ldots, x_n$ is

$$\bar{x} = (x_1 + x_2 + x_3 + \cdots + x_n)/n$$

or, more succinctly,

$$\bar{x} = \left(\sum_{i=1}^{n} x_i\right)/n. \tag{4.1}$$

Using the gene expression data in Table 4.1 we obtain the mean of the three values for Gene 1 under the low-calcium challenge to the resistant strain as $\bar{x} = 6.326$. Similarly, for Gene 2 under the low-calcium challenge to the wild-type strain the mean is 6.180. These two values are very close, but looking at the two sets of results we can see a clear difference between the two. For Gene 1, the three individual values differ from the mean by $-1.118, 0.323$ and 0.796, whereas for Gene 2 they differ from the mean by $0.177, 0.098$ and -0.274. We say that the results are more *variable* for Gene 1 than for Gene 2.

The average of each set of three differences is necessarily zero, but we can obtain a measure of variability by averaging the squared differences. This gives the *sample variance*, whose definition is

$$s^2 = \left\{\sum_{i=1}^{n}(x_i - \bar{x})^2\right\}/(n-1). \tag{4.2}$$

Note that this is not quite an average, because the divisor in (4.2) is not n, but $n-1$. This makes little difference in large samples, but gives a more reliable answer in small samples and, incidentally, acknowledges that you can't measure variability using a sample of size 1. Note also that the sample variance does not have the same physical dimensions as the original data values; for example, the units of the sample variance of a set of distances has dimension distance-squared. To restore the original physical dimensions, we take a square root, to give the *sample standard deviation*,

$$SD_x = \sqrt{\{\sum_{i=1}^{n}(x_i - \bar{x})^2\}/(n-1)} \tag{4.3}$$

Table 4.3. Sample means (upper rows of each pair) and standard deviations (lower rows of each pair) of pre-processed gene expression data for four randomly selected genes in the *Arabadopsis* experiment: 1 = 253023.at, 2 = 244928.s.at, 3 = 262200.at, 4 = 250740.at.

Treatment		Gene			
Strain	Ca challenge	1	2	3	4
Ca resistant	low	6.326	6.708	15.395	9.064
		0.997	0.326	2.555	1.187
	high	5.280	7.714	15.865	9.748
		0.684	0.716	0.768	2.174
wild-type	low	5.691	6.180	17.471	8.683
		0.641	0.241	6.299	0.264
	high	5.828	6.205	17.458	9.693
		0.784	2.081	2.795	3.098

The means and standard deviations for all 16 sets of three replicates in Table 4.1 give the summary in Table 4.3.

A summary table is hardly necessary with only three replicates, but for larger batches direct inspection of individual values becomes difficult or impossible, whereas means and standard deviations are still readily interpretable.

A slightly less compact, but correspondingly more informative, summary of a single batch of data is the *five-number summary*: (minimum, lower quartile, median, upper quartile, maximum). This summary requires the data values to be ordered from smallest to largest, then the minimum and maximum are at the two extremes of the ordered sequence, the quartiles are one-quarter and three-quarters of the way along, and the median is half way along. The median gives an alternative to the sample mean as a measure of average value, whilst the difference between the two quartiles, called the *inter-quartile range* is an alternative to the standard deviation as a measure of variability. The minimum and maximum are useful for checking that the data do not include gross outliers that might arise through coding errors. The five-number summary also conveys information about distributional shape, as we now demonstrate through an example.

The 25 ordered values of log-base-two-transformed gene expression levels shown in Figure 4.3 are:

2.231 2.262 2.366 2.387   2.443   2.501 2.954 3.004 3.021 3.138
3.404 3.408 4.229 4.255   4.337   4.347 4.803 4.843 5.677 6.168
6.220 6.687 6.718 7.700 11.261

The five-number summary for these data is

$$2.231 \quad 2.954 \quad 4.229 \quad 5.677 \quad 11.261$$

The inter-quartile range is $5.677 - 2.954 = 2.723$. Note that the upper quartile is further above the median than the lower quartile is below the median, and that the maximum is further above the upper quartile than the minimum is below the lower quartile. Both of these features indicate a positively skewed distributional shape, as we have already seen in Figures 4.3 and 4.4.

4.7.3 *Summarizing relationships*

We now revisit the data shown in Table 4.2. Figure 4.11 showed a scatterplot of the paired expression levels of genes A and B, from which we concluded that 'there is perhaps a hint of a positive association between the two genes'. How could we express this numerically? The standard way of doing so is by the *sample correlation*,

$$r = \frac{\sum_{i=1}^{n}(x_i - \bar{x})(y_i - \bar{y})/(n-1)}{SD_x SD_y}, \qquad (4.4)$$

where x and y denote the two values in each pair (here, expression levels for genes A and B, respectively), and n is the sample size (here, $n = 12$). For the data in Table 4.2 this gives $r = 0.303$.

Why is the sample correlation a measure of association? The numerator on the right-hand side of equation (4.4) is an average of the product of two quantities, each of which necessarily has an average value zero, so the products will in general contain a mixture of positive and negative values. If x and y are positively associated, then values of x greater than $\bar{x}$ will tend to be paired with values of y greater than $\bar{y}$, and similarly for values less than their respective sample means. This leads to a preponderance of positive products and hence a positive value of r. If x and y are negatively associated, values of x greater than $\bar{x}$ will tend to be paired with values of y less than $\bar{y}$ and vice versa, there will be a preponderance of negative products, and r will be negative. Finally, the denominator on the right-hand side of equation (4.4) serves two purposes: it makes r a dimensionless quantity so that the correlation between two variables does not depend on the units of measurement; and it constrains r to lie between -1 and $+1$. Perfect correlation ($r = \pm 1$) is only obtained when the pairs (x, y) fall exactly along a straight line in their scatterplot. Zero correlation indicates

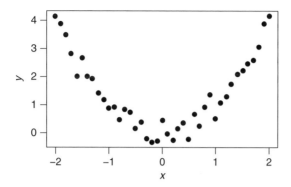

Fig. 4.22. Synthetic data showing a strong non-linear relationship between x and y, but a sample correlation close to zero ($r = -0.025$).

a complete absence of *linear* association. The emphasis here is important. A single number cannot capture the rich variety of possible associations between pairs of measurements. For example, Figure 4.22 shows a synthetic dataset for which the sample correlation is $r = -0.025$, yet there is a very strong, albeit non-linear, relationship between the two variables. Describing non-linear relationships in numerical terms requires statistical modelling of the relationship, a topic we discuss in Chapter 7.

The other widely used measure of association is between two binary variables. Suppose, for example, that we want to establish whether the presence of a specific genetic variant is associated with the risk of a particular disease. One way to do this is to obtain a sample of *cases* of the disease and a second sample of people who do not have the disease, called *controls*. For each person we then determine whether they do or do not have the genetic variant. The resulting data can be presented in a two-by-two table, an example of which is the following:

	genetic variant	
	absent	present
cases	85	15
controls	183	17

The proportions of people with the genetic variant differ between cases and controls, $15/100 = 0.15$ and $17/200 = 0.085$, respectively. This suggests a positive association between the genetic variant and the disease, which we can express in a single number, the *odds ratio*,

$$OR = \frac{15/85}{17/183} = 1.90.$$

Had the two proportions been identical, we would have obtained $OR = 1$. The observed value $OR = 1.90$ implies a near-doubling of the risk of disease when the genetic variant is present. Values of OR less than one indicate negative association.

5
Experimental design: agricultural field experiments and clinical trials

5.1 Agricultural field experiments

Rothamsted agricultural research station, located in the UK county of Hertfordshire, has been operating since 1843 and is 'almost certainly the oldest agricultural research station in the world' (http://www.rothamsted.ac.uk/). Throughout Rothamsted's history, its scientific staff have conducted field experiments to compare crop yields, either of different varieties of a particular species, or of different experimental treatments applied to the crops, for example, different combinations of soil fertilizer. An agricultural field trial involves selecting a suitable piece of land on which to carry out the experiment, subdividing the land into *plots*, allocating one of the varieties or *treatments* to each plot, and subsequently measuring the *yield* from each plot. In the terminology of Chapter 2, the treatment allocation is the *input* variable and the yield the *output* variable.

Figure 5.1 shows an aerial view of part of the Rothamsted site, on which a field trial is being conducted. The plots are rectangular in shape, and their boundaries are clearly visible. In experiments of this kind, the size and shape of each plot is usually constrained by practical considerations, specifically the need to sow and harvest individual plots without interference across the boundaries; sometimes, the design will include buffer strips running between the plots themselves to minimize interference effects such as the leaching of fertilizer across plot boundaries. From a statistical viewpoint, two key questions are: how many plots should we use? and how should we allocate the different experimental treatments amongst the plots?

The answer to the first of these questions necessarily involves a balance between precision and cost. If similar experiments, or pilot studies, have been performed previously their results should give some indication of the likely magnitude of the random variation between yields from plots that receive the same treatment. Otherwise, the choice will usually be dictated

Fig. 5.1. Part of the Rothamsted research station, showing field trials in progress. This figure is reproduced in colour in the colour plate section.

by the available resources. Sometimes, so-called 'power calculations' are made to determine how many plots are needed to achieve the experiment's stated objectives. We will discuss power calculations later in this chapter, in Section 5.6.

Careless allocation of treatments to plots can easily result in a worthless experiment. For example, most agricultural land is, to a greater or lesser extent, heterogeneous in character because of spatial variation in the local micro-environment, including soil fertility, aspect, slope and any other environmental factors that might influence crop yield. This is well illustrated by Figure 5.2, which shows results obtained in a special type of experiment, called a *uniformity trial*. In this experiment, 500 rectangular plots were arranged in a 20 row by 25 column array. Each plot received exactly the same treatment, hence the results show how much variation in yield might typically result from variation in the micro-environment; the experiment was a uniformity trial of wheat and was carried out at Rothamsted early in the twentieth century; for details, see Mercer and Hall (1911). The resulting plot yields show a very clear spatial trend so that, for example, had this been a comparative trial of different varieties of wheat, any variety planted

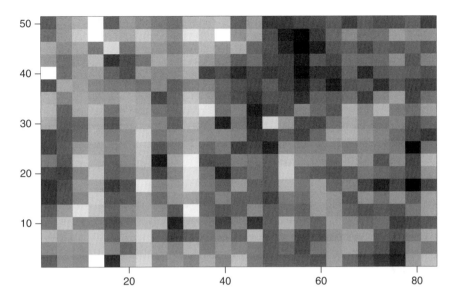

Fig. 5.2. Image-plot of wheat yields in a uniformity trial. The colour coding runs from black (low) through grey to white (high).

predominantly towards the left-hand side of the field would have been at an unfair advantage.

At the time of the Mercer and Hall experiment, statistical science was in its infancy and systematic methods for designing and analysing field trials to best advantage were unknown. The statistician R. A. Fisher (1890–1962) was employed at Rothamsted between 1919 and 1933. During this period he revolutionized the theory and practice of statistics, especially as it applies to agricultural experimentation (especially, but not exclusively—Fisher is revered not only as a great statistician but also as one of the great geneticists of the twentieth century). Figure 5.3 shows Fisher working on the mechanical calculator which was the only computing machine available to him. Box (1978) is an excellent scientific biography of Fisher.

Two of Fisher's fundamental contributions to experimental design were the ideas of *randomization* and *blocking*, which we now discuss in turn.

5.2 Randomization

Figure 5.2 suggests strongly that the allocation of treatments to plots can materially affect the results of an experiment. Suppose, for example, that the Mercer and Hall experiment had not been a uniformity trial, but a comparative trial of two varieties of wheat, with one variety planted on the left-hand half of the field (columns 1 to 12 and half of column 13), the other on the right-hand side. Figure 5.4 shows the distributions of yields from the two

Fig. 5.3. R. A. Fisher at work.

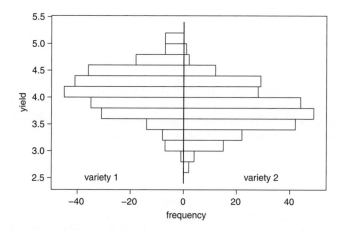

Fig. 5.4. Distributions of wheat yields from the Mercer and Hall uniformity trial. Yields from the left-hand side of the field are shown to the left of the y-axis, yields from the right-hand side of the field to the right of the y-axis.

fictitious varieties of wheat, and gives a visual impression that variety 1 tends to produce higher yields, on average, than variety 2; the two sample means are 4.116 and 3.768. But how can we be confident that this represents something more than random variation between identically treated plots? To answer this question we introduce the idea of a *standard error*.

In equation (4.3) we defined the standard deviation, SD, as a measure of the variability in a single batch of data. The more variable the data, the more imprecise must be our estimate of the mean of the population from which the data were drawn. But however large or small the variability in the population, we can obtain a more precise estimate by increasing the sample size, n. The mathematically correct way to combine these two considerations into a measure of the precision of a sample mean is the *standard error*, $SE = SD/\sqrt{n}$.

In this example, the two sample means and standard errors are 4.116 and 0.028 for variety 1, compared with 3.7681 and 0.026 for variety 2. A superficial analysis would suggest that variety 1 is superior to variety 2. But we know this to be false, because in reality there was only one variety.

Fisher's solution to this problem was to recommend that the allocation of treatments to plots should always be made at random. We illustrate this for the Mercer and Hall data by randomly selecting 250 of the 500 plots to receive the fictitious variety 1, and the remainder to receive the equally fictitious variety 2. Figure 5.5 shows the resulting distributions of yields from the two varieties; means and standard errors for the yields were 3.926 and 0.027 for variety 1, compared with 3.971 and 0.031 for variety 2. A reasonable, and now correct, conclusion is that there is no difference between the two varieties.

An obvious virtue of randomization is that it eliminates any possibility of subjective bias in the allocation of treatments amongst plots. Fisher argued more strongly than this, to the effect that randomization provides a basis for carrying out a formal test of the hypothesis that any apparent difference amongst the treatments can be explained entirely by chance, i.e., that the treatments themselves have no effect on yield. We again use the Mercer and Hall data to illustrate this idea. Note firstly that in our previous illustration we observed that variety 1 gave an average yield of 3.926, smaller than the average yield of 3.971 for variety 2. The difference between the two averages is 0.045 in favour of variety 2. In the previous paragraph, we suggested that this was consistent with what we know to

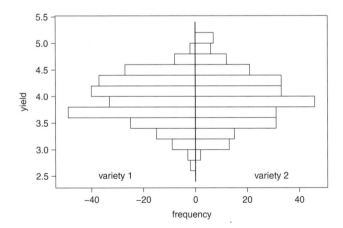

Fig. 5.5. Distributions of wheat yields from the Mercer and Hall uniformity trial. Yields from a random sample of 250 of the 500 plots are shown to the left of the y-axis, yields from the remaining 250 plots to the right of the y-axis.

be true, namely that the two varieties are in fact identical, but we gave no formal justification for this suggestion. Suppose that we now randomly permute the 500 observed yields over the 500 plots and recalculate the difference between the average yields for the two varieties. If the varieties are identical, the original randomization of varieties amongst plots and the random permutation of observed yields amongst plots are equivalent processes. Furthermore, we can repeat the random permutation as often as we like, and so build up a picture of how variable the difference between the two average yields might be under the hypothesis that the two varieties are identical. The left-hand panel of Figure 5.6 shows the result after 999 random permutations. The histogram gives the distribution of the differences between the two average yields over the 999 random permutations, whilst the solid dot shows the original difference of 0.045. By way of contrast, the right-hand panel repeats this exercise, but after first adding 0.1 to the yield of each plot that received variety 2 in the original allocation. The two panels tell very different stories. In the left-hand panel, the solid dot is a perfectly reasonable value from the distribution represented by the histogram, whilst in the right-hand panel it is not. This is as it should be, since we know that in the right-hand panel there is indeed a genuine difference of 0.1 between the average yields of the two varieties. In Chapter 6 we will show how this idea can be set within the formal framework of statistical hypothesis testing.

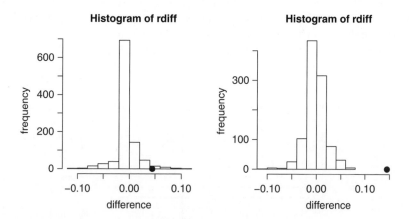

Fig. 5.6. Illustration of a completely randomized design based on the Mercer and Hall uniformity trial data. In the left-hand panel, the 500 wheat yields were randomly split into two sets of 250, and each set assigned to one of two (fictitious) varieties. The solid dot shows the value of the difference between the average yields for the two sets of 250 values, whilst the histogram shows the distribution of the difference between the two average yields when the random split into two sets is repeated 999 times. In the right-hand panel, the same procedure was followed, except that 0.1 was added to each of the 250 yields originally assigned to variety 2.

Fisher did not have a computer available to him, and would have had to construct each random permutation by physically drawing 250 tickets at random from a set of 500. Instead, he used his mathematical skill to work out the theoretical form of the statistical distribution induced by random permutation.

A design of this kind is called a *completely randomized design*.

5.3 Blocking

The primary purpose of randomization is to ensure honesty on the part of the experimenter. Randomization enables valid inferences to be made from the results of the experiment, but does not guarantee that the experiment has been designed as efficiently as possible. What do we mean by this? Again taking the Mercer and Hall data as an illustration, suppose now that the experiment involved a comparison between 20 different wheat varieties, with the aim of comparing their average yields. Once again, a random allocation of varieties amongst the 500 plots will allow an honest comparison. Table 5.1 shows a selection of observed average differences between pairs of varieties along with the corresponding standard errors. In this illustration, we have created some genuine differences amongst the varieties by adding different constants to the yields according to their

Table 5.1. Pairwise comparisons between selected treatments (varieties) in a simulation of a *completely randomized* experiment based on the Mercer and Hall data, with 20 treatments and 25 plots allocated at random to each treatment. The three columns in each row give the labels of the two treatments being compared, the difference between the average of 25 observed yields, and the standard error of this difference.

Comparison	Mean difference	Standard error
1 vs. 2	1.917	0.111
1 vs. 3	0.024	0.115
1 vs. 4	2.859	0.125
1 vs. 5	−0.130	0.109
2 vs. 3	−1.893	0.125
2 vs. 4	0.942	0.134
2 vs. 5	−2.046	0.119
3 vs. 4	2.836	0.137
3 vs. 5	−0.153	0.122
4 vs. 5	−2.989	0.132

treatment allocations. The sizes of the observed average differences in relation to their standard errors now allows us to assess, after the event, the likely magnitude of these differences. A conventional way to summarize the true difference in average yields between any two varieties is as the observed mean difference plus and minus two standard errors. For varieties 1 and 2 this gives the range $(1.695, 2.139)$. Why this is a reasonable convention will be explained in Chapter 6. Note also that the average of the standard errors is 0.123, which we can take as a summary measure of the experiment's precision.

The observed variation in yields has at least two components: environmental variation amongst the 500 plots; and genetic variation in seed quality. We can do nothing about the second of these, but we can reduce the first by insisting that comparisons between different varieties are made amongst environmentally similar subsets of plots. In general, to identify such subsets we would need to know more about the field in question. However, a good general rule, sometimes called the first law of geography, is that spatially adjacent plots are more likely to be similar than spatially remote plots. To exploit this, we now design the experiment somewhat differently. Each of the 25 columns of the rectangular array consists of 20 plots. Rather than randomly allocate 25 of the 500 plots to receive each of the 20 treatments, we randomly allocate each treatment to exactly one plot in each column. Now, to compare two treatments we calculate the observed difference in the two yields within each column, and average the result. Table 5.2 shows a selection of the resulting comparisons, in the same format as Table 5.1. The average differences between pairs of treatments are similar in both tables, but in most cases the associated standard error is smaller in Table 5.2 than in Table 5.1, and the average of the standards errors has decreased from 0.123 to 0.099. The explanation for this is that by comparing treatments within each column, we have eliminated some of the environmental variation from the treatment comparisons of interest and so obtained a more precise set of results.

Fisher coined the term *block* to mean a subset of plots which, a priori, would be expected to be relatively similar in character. The experimental design that we have just illustrated is called a *complete randomized block design* with 20 treatments in 25 blocks, where 'complete' refers to the fact that every treatment appears in every block. In many experimental settings, we cannot guarantee that the number of plots per block is equal to the number of treatments, and this gives rise to various *incomplete* randomized block designs.

The effectiveness of blocking as a strategy to increase the precision of an experiment depends on the amount of environmental heterogeneity amongst the plots and on the experimenter's ability to identify in advance relatively homogeneous subsets of plots to define the blocks.

Table 5.2. Pairwise comparisons between selected treatments in a simulation of a *complete randomized block* experiment based on the Mercer and Hall data, with 20 treatments in 25 blocks. The three columns in each row give the labels of the two treatments being compared, the average of the 25 pairwise differences in observed yields, and the standard error of this difference.

Comparison	Mean difference	Standard error
1 vs. 2	1.771	0.113
1 vs. 3	−0.077	0.100
1 vs. 4	2.930	0.098
1 vs. 5	−0.076	0.080
2 vs. 3	−1.848	0.101
2 vs. 4	1.159	0.110
2 vs. 5	−1.846	0.102
3 vs. 4	3.007	0.112
3 vs. 5	0.001	0.107
4 vs. 5	−3.006	0.064

5.4 Factorial experiments

A factorial experiment is one in which the treatments under investigation have an internal structure. For example, in the *Arabadopsis* experiment described in Section 4.1 the four treatments were made up of all combinations of strain (calcium resistant or wild-type) and calcium challenge (low or high). Strain and calcium challenge are *factors*, each of which takes one of two *levels*. This defines a 2×2 *factorial* treatment structure. The definition extends in a natural way to factorial structures with more than two factors at more than two levels. Thus, for example, the data in Table 4.1 can now be seen as three replicates of a $2 \times 2 \times 4$ factorial whose factors are strain, calcium challenge and gene, respectively.

Factorial experiments enable the investigation of *interactions* between factors. In the *Arabadopsis* experiment, two factors are said to interact if the effect of changing the level of one factor depends on the level of the other factor. Table 5.3 shows the sample means of log-base-two transformed expression level observed for Gene 2 in each of the four treatments, together with row and column means whose interpretation we now discuss.

The difference between mean log-transformed gene expression levels in the wild-type and Ca-resistant strains, averaged over the two levels of calcium challenge, is $2.844 − 2.598 = 0.246$, with the Ca-resistant strain

Table 5.3. Observed means of log-base-two-transformed expression levels for Gene 2 in Table 4.1.

	Ca challenge		
Strain	Low	High	Mean
Ca resistant	2.745	2.943	2.844
wild type	2.627	2.570	2.598
mean	2.686	2.756	2.721

giving the higher value. This is the estimated *main effect* of the strain factor. Similarly, the estimated main effect of the calcium challenge factor is $2.756 - 2.686 = 0.070$, with the high calcium challenge giving the higher value. We can calculate the interaction between the two factors in two equivalent ways. The first is to estimate the effect of strain at the high level of calcium challenge, $2.943 - 2.570 = 0.373$, and at the low level of calcium challenge, $2.745 - 2.627 = 0.118$. Then, the estimated *interaction* is the difference between the two, $0.373 - 0.118 = 0.255$. Interchanging the roles of the two factors in this calculation would give the same answer. Note that the ordering of the two levels of each factor is arbitrary, but in calculating the interaction effect the same ordering must be used for both parts of the calculation. Fortunately, statistical software handles this kind of thing for us automatically.

The concept of interaction is related to whether the effects of the two factors concerned are *additive*, i.e., the effect of changing from the first to the second level of both factors is the *sum* of the separate effects of changing from the first to the second level of each factor separately. The data from the *Arabadopsis* experiment are being analysed after a log-base-two transformation of each measured gene expression level, meaning that an additive effect is equivalent to a *multiplicative* effect on the original scale. If we repeat the previous calculation on untransformed expression levels, the estimated interaction becomes 0.981. Whether either of these figures represents a genuine effect rather than random sampling fluctuation in the data is a question we postpone until Chapter 7.

When there is no interaction between a pair of factors, a factorial design is more efficient than a pair of single-factor experiments. Suppose, for example, that the 12 arrays in the *Arabadopsis* experiment had been assigned to treatments as follows: four to the Ca-resistant strain under the low Ca challenge; four to the wild-type strain under the low Ca challenge; four to the Ca-resistant strain under the high Ca challenge. Then, a comparison between arrays 1 to 4 and 5 to 8 estimates the strain effect and a comparison between arrays 1 to 4 and 9 to 12 estimates the Ca-challenge effect, i.e., in each case 8 arrays contribute to the estimate of each effect. If, instead, we allocated three arrays to each of the four treatments as was in fact done, then a comparison between arrays 1 to 3 and 4 to 6 estimates the Ca-challenge effect, but so does a comparison between arrays 7 to 9 and 10 to 12, i.e., all 12 arrays contribute to the estimation of the Ca-challenge effect and, by the same argument, to the estimation of the strain effect. The result is to increase the precision of each estimate by a factor $\sqrt{12/8} = 1.22$.

5.5 Clinical trials

In the medical and health sciences, experiments to compare the effectiveness of two or more treatments are called *clinical trials*. Although this terminology is specific to medical settings, the principles used to design clinical trials are exactly the same as those described above in the context of agricultural field trials, but with the added complication that ethical considerations are also paramount when conducting experiments on people, rather than on fields of wheat.

Fisher's original work on the design and analysis of agricultural experiments emphasized the importance of randomization as a means of ensuring the validity of inferences concerning differences amongst the treatments being compared. The pioneer advocate of randomization in the context of medical research was Sir Austin Bradford Hill (1897–1991). His work emphasized the importance of randomization as a way of protecting against conscious or unconscious bias in the allocation of patients to treatments.

One of the key ethical considerations in a clinical trial is that a physician cannot agree to treat a patient using an intervention that they know, or strongly believe, to be inferior to the best available option. The first example of a modern clinical trial was an investigation into the effectiveness of streptomycin, an early antibiotic, in the treatment of tuberculosis (Medical Research Council, 1948). At the time, streptomycin was in short supply, and this provided an ethical argument in favour of randomization. As there were good grounds for thinking that streptomycin would be effective, its scarceness reversed the ethical argument to one in favour of randomization: if not all patients could receive an intervention that was highly likely to prove beneficial, the only ethically acceptable way to determine who would be treated was to draw lots.

The present-day resolution of these ethical issues relies on the concept of *equipoise* and on the practice of *blinding*. Equipoise requires that a trial is only carried out when there is genuine doubt as to which, if any, of the interventions being compared are superior to others included in the trial. *Blinding* requires that neither the patient nor their treating physician know to which intervention the patient has been randomized. An immediate corollary is that all patients in a trial must give their *informed consent* to take part, implying amongst other things, their acceptance that they cannot choose a particular intervention.

A fascinating set of reflections on the early history of clinical trials, and in particular the role of randomization, can be found in Armitage *et al.* (2003). Clinical trials methodology has since grown into an academic discipline in its own right, covering a wide range of statistical and non-statistical considerations. A detailed account is given by Piantadosi (2005).

5.6 Statistical significance and statistical power

In Section 2.7 we described briefly three kinds of statistical inference, namely parameter estimation, hypothesis testing and prediction. There is increasing recognition amongst statisticians that hypothesis testing tends to be overused, and parameter estimation underused, in many areas of scientific work. However, it continues to play a key role in the analysis of data from clinical trials. This is because a clinical trial is typically conducted to obtain an unequivocal answer to a specific question, for example to make a decision whether or not to licence a novel drug compound for medical use, rather than to contribute to the general understanding of a possibly complex scientific process. We shall therefore use the context of a simple comparative trial to discuss the formal framework of hypothesis testing.

Suppose that we wish to compare a novel treatment for a particular medical condition with the currently recommended standard treatment. The *null hypothesis* to be tested is that there is no difference in efficacy between the two treatments. The *alternative hypothesis* is that the novel treatment is superior. The standard protocol for a clinical trial to answer this question is the following (admitting some simplification): define the population of eligible patients (for example, all people diagnosed with the medical condition, say chronic hypertension); decide how to measure the efficacy of treatment, called the *primary outcome* (for example, average systolic blood pressure measured daily for seven days, two weeks after beginning treatment with either the standard or novel treatment); recruit a sample of patients from the eligible population; randomly allocate either the standard or the novel treatment to each patient; determine the value of the primary outcome for each patient; analyse the data on the primary outcome and decide whether there is or is not convincing evidence to justify rejection of the null hypothesis in favour of the alternative hypothesis.

For our hypothetical example, assume that n patients have been allocated to each of the two treatments, and let $\bar{X}$ and $\bar{Y}$ denote the sample mean of the primary outcome for patients allocated to the standard and novel treatments, respectively. It would seem reasonable to *reject* the null hypothesis if $T = \bar{Y} - \bar{X}$ is sufficiently large, i.e., $T > c$ for some constant c. The value of c that we choose will determine the probability that we will reject the null hypothesis, $\mathcal{H}_0$, when it is in fact true. Write this as

$\alpha = P(T > c|\mathcal{H}_0)$, to be read as 'the probability that T will be greater than c given that the null hypothesis is true'. Since this represents a mistake, we would like this probability to be small, which we can easily do by making c very large. But this would be unhelpful, because it would imply that we would be unlikely to reject $\mathcal{H}_0$ even when $\mathcal{H}_0$ is false. For example, choosing $c = 25$ mm Hg would risk missing a potentially valuable clinical improvement in blood pressure from using the novel rather than the standard treatment. The conventional way round this dilemma is to choose c such that the probability of rejecting a true null hypothesis is controlled at a pre-specified value, for example, $\alpha = 0.05$, called the *prescribed significance level* of the test and usually expressed as a percentage rather than a probability. In order to control the behaviour of the test in this way, the value of c will depend on the number of patients recruited to the trial, which we indicate by writing it as c_n.

Now recall that the result of a hypothesis test can be wrong in either of two ways. By prescribing the significance level we have controlled the probability of making one kind of mistake, namely rejecting a true null hypothesis, but this says nothing about the probability of rejecting a false null hypothesis. Write this second probability as β. If the novel treatment is materially better than the standard treatment, we would like β to be as close to 1 as possible. The difference between the population mean values of the primary outcome under the novel and standard treatments is called the *clinically significant* difference, or in statistical terminology, the *alternative hypothesis*, whilst the corresponding value of β is called the *power* of the test. Writing $\mathcal{H}_1$ for the alternative hypothesis, we arrive at the scenario summarized by the following table of probabilities:

Decision	True state of nature	
	$\mathcal{H}_0$	$\mathcal{H}_1$
accept $\mathcal{H}_0$	$1 - \alpha$	$1 - \beta$
reject $\mathcal{H}_0$	α	β

Now, for prescribed values of α and the clinically significant difference, $\mathcal{H}_1$, the value of β will depend on the sample size, n. This is intuitively reasonable: the more data you have, the more clearly you should be able to see whether or not the novel treatment is better than the standard treatment. However, there are two reasons for not driving β arbitrarily close to one by increasing the sample size indefinitely. The obvious one is that the larger the sample size the greater the cost. The more subtle one is that if it becomes reasonably clear that the novel treatment is materially better than the standard treatment, it is then unethical to continue to allocate patients to an inferior treatment. Just as 5% has become the accepted standard for statistical significance, so 80% has become the accepted, if somewhat less entrenched, standard of power to detect a clinically significant difference. The term *power calculation* or, more accurately, *sample size calculation*, is used to denote the calculation of the sample size necessary to achieve both the prescribed significance level and the required power to detect the clinically significant difference.

In our experience of designing clinical trials, it is often unclear exactly what threshold should be used to define clinical significance, there is imperfect knowledge of the amount by which the primary outcome will vary between identically treated patients, and there are practical constraints on the number of patients who can be recruited. For this reason, we always carry out power calculations under a range of scenarios, which we then discuss with our clinical collaborators. And we always ask whether a power calculation is really what is needed, as opposed to a calculation of how precisely we can estimate the size of the difference in efficacy between the treatments concerned.

5.7 Observational studies

In some branches of science, designed experiments are the exception rather than the rule. Instead, the prevailing paradigm is to observe nature in its natural state. Investigations of this kind are called *observational studies*. In general, these tend to be subject to more sources of extraneous variation than do designed experiments and this has two consequences. Firstly, they tend to yield less precise results. Secondly, it is difficult to specify all the details of the analysis protocol in advance of the investigation; the approach to data analysis often needs to be more flexible, using the ideas of statistical modelling that feature in Chapter 7.

6
Simple comparative experiments: comparing drug treatments for chronic asthmatics

6.1 Drug treatments for asthma

Asthma is a chronic condition whose characteristic symptom is an abnormal degree of wheezing or breathlessness. Typically, any one patient's severity of symptoms ebbs and flows over time, this being in part a response to changes in ambient air quality related to weather conditions or exposure to air-borne allergens such as pollen.

Anti-congestant drugs are intended to relieve the acute symptoms of asthma. Typically, an asthmatic patient will self-administer their drug by inhaling from a small vaporizing device that releases a metered dose, either as an emergency response to an acute attack or as part of a regular schedule, for example, before going to bed each night. The effectiveness of the drug can be assessed by measuring the maximum rate at which the patient is able to exhale, denoted by PEF (Peak Expiratory Flow), a fixed time after administration of the drug.

A clinical trial to compare two anti-congestant drugs, Formoterol and Salbutamol was conducted some years ago, and reported by Graff-Lonnevig and Browaldh (1990). The results from the trial are shown in Table 6.1.

In the remainder of this chapter, we will discuss a number of ways in which the trial might have been designed (including the one that was actually used), and in each case how the data would have been analysed.

6.2 Comparing two treatments: parallel group and paired designs

The asthma trial described above is an example of a simple comparative experiment, whose defining features are the following. Each replicate of the experiment generates a single measure, called a *response*. Replicates can be performed under each of several qualitatively different experimental

Table 6.1. Values of PEF recorded in a clinical trial to compare the effectiveness of two anti-congestant drug treatments, Formoterol (F) and Salbutamol (S), in the relief of acute asthmatic symptoms (data kindly provided by Professor Stephen Senn).

Drug F	Drug S
310	270
310	260
370	300
410	390
250	210
380	350
330	365
385	370
400	310
410	380
320	290
340	260
220	90

conditions which collectively define a single design variable, or *treatment*. The objective of the experiment is to compare the mean responses under the different treatments.

In this section, we further simplify to the case of two treatments; in the asthma study, these correspond to the two drugs, Formoterol and Salbutamol.

6.2.1 *The parallel group design*

The most straightforward design for a simple comparative study of two treatments is the parallel group design. In this design, if a total of n experimental units are available, we choose m of these completely at random to receive treatment A, say, and the remaining $n - m$ then receive treatment B. Except in special circumstance, for example if there are big differences between the costs of applying the two treatments, the most efficient choice for fixed n is to take $m = n/2$, i.e., equal numbers of units allocated to the two treatments.

The parallel group design is appropriate when we have no reason a priori to distinguish between different experimental units. The design is very widely used in clinical trials where the unit is the patient; patients are recruited one at a time and ethical considerations require that every patient is afforded equal access to each of the treatments on offer.

Setting ethical considerations aside for the moment, one consequence of using a parallel group design is that all of the inherent variation between patients contributes to the uncertainty in the results obtained; by design, variation between patients is treated as random variation, and if this variation is relatively large, so will be the standard error attached to the estimated difference between the mean response under the two treatments.

6.2.2 *The paired design*

The paired design is appropriate when there is a natural way of arranging the experimental units into pairs so that we might expect results from the two units within a pair to be more similar than those from two units in different pairs. Pairing is a special case of blocking, in which the block size is two. In the context of the asthma trial, an extreme form of pairing is available, because asthma is a chronic condition and the treatments being compared are intended only to give short-term relief from symptoms, rather than to effect a long-term cure, i.e., each child in the study can be given both treatments, one after the other, with a sufficient gap between the two that no residual effect of the first drug remains when the second drug is administered. Randomization still has a role to play in the design because the severity of a patient's asthmatic symptoms varies according to various external factors, including air quality and weather conditions. Hence, the order in which each patient receives the two drugs can be chosen at random, e.g., by the flip of a fair coin. The advantage of the pairing is that differences between the two drugs are no longer compounded by differences between children. How much of an advantage this is depends on the context.

6.3 Analysing data from a simple comparative trial

6.3.1 *Paired design*

Although the paired design is conceptually more sophisticated, than the parallel group design, its analysis is easier. So we shall consider it first.

Suppose (as was in fact the case) that the two PEF measurements in each column of Table 6.1 were from the same patient, i.e., the design was a paired design. The first step in the analysis of the data is, as always, to display the results graphically to check for any unusual features: perhaps an unnecessary precaution for such a small dataset where it is easy to eyeball the data, but a wise one otherwise. Figure 6.1 shows the values of F and S as a scatterplot, exploiting their pairing. This confirms that the paired values of F and S are positively correlated and thereby demonstrates the value of the paired design: the concentration of the plotted points along the diagonal implies that the differences between paired values of F and

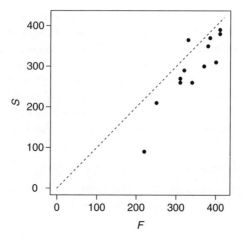

Fig. 6.1. Scatterplot of PEF values recorded on each of 13 asthmatic children when treated with Formoterol (F) or Salbutamol (S).

S are less variable than either F or S individually. The plot also shows that the extreme values $F = 220$ and $S = 90$ are, from a purely statistical perspective, not grossly inconsistent with the general pattern of variation and may therefore indicate that one of the patients is a particularly severe case. Had this patient given an extreme value for one of the drugs and a typical value for the other, we might have questioned the accuracy of the extreme value.

Having accepted the data as they are, the next step is to calculate the differences, $d = F - S$ say, between the two results for each patent. This gives:

$$40 \quad 50 \quad 70 \quad 20 \quad 40 \quad 30 \quad -35 \quad 15 \quad 90 \quad 30 \quad 30 \quad 80 \quad 130$$

The next step is to calculate the average value of d, namely

$$\bar{d} = 45.4,$$

which would seem to suggest that F is the more effective drug. But to be reasonably confident that this is not a chance finding, we need to assess the variability in the data, and hence the precision of $\bar{d}$ as an estimate of how much more effective on average F might be than S if prescribed routinely.

The final step is therefore to calculate the standard error of our estimate,

$$SE(\bar{d}) = SD/\sqrt{n}, \tag{6.1}$$

where $SD = 40.59$ is the standard deviation of the values of D and $n = 13$ is the sample size, i.e., the number of pairs of measured values of PEF. We then report an approximate 95% *confidence interval* for the mean difference in the effectiveness of the two drugs as

ANALYSING DATA FROM A SIMPLE COMPARATIVE TRIAL

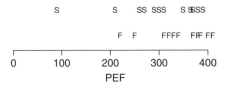

Fig. 6.2. Dot-plot of unpaired PEF values for each of 13 children when treated with Formoterol (F) and with Salbutamol (S).

$$\bar{d} \pm 2 \times SE(\bar{d}) = (22.9, 67.9).$$

Table 6.2 summarizes the analysis in a form adaptable to any paired design, or to the calculation of a 95% confidence interval for the population mean of a single batch of data. The name refers to the fact that intervals constructed in this way have a 95% chance of including the true, but unknown, difference between the two population means, i.e. the true difference in efficacy.

Health warning. The number '2' in the last line of Table 6.2 should strictly be replaced by a number that depends on the sample size, n, but is within ±0.05 of 2 for n bigger than 27. This need not concern you if you use statistical software packages to calculate confidence intervals, but it does mean that you may get slightly different answers to those quoted above.

6.3.2 Parallel group design

Had the data in Table 6.1 been derived from a parallel group trial, the scatterplot shown in Figure 6.1 would have made no sense, because the alignment of the columns would have been arbitrary. An appropriate diagram would then have been the dot-plot shown as Figure 6.2.

The interpretation of Figure 6.2 is not entirely clear. There is a suggestion that F is more effective on average than S, but the results show wide

Table 6.2. Stages in the calculation of a 95% confidence interval for the population mean of a single batch of data with sample size n.

Data	$x_1, x_2, \ldots, x_n$
Sample mean	$\bar{x} = n^{-1} \sum_{i=1}^{n} x_i$
Sample variance	$s^2 = (n-1)^{-1} \sum_{i=1}^{n} (x_i - \bar{x})^2$
Standard error	$SE(\bar{x}) = \sqrt{s^2/n} = SD/\sqrt{n}$
95% confidence interval	$\bar{x} \pm 2 \times SE(\bar{x})$

variation and considerable overlap between the two drugs. Also, without the pairing, the extreme result $S = 90$ does looks a bit odd. If we again accept the data as they stand, the analysis now begins by calculating separate sample means, variances and standard deviations for each of the two sets of results. This gives

$$\bar{F} = 341.2 \quad s_F^2 = 3559.32 \quad SD = 59.66 \quad \bar{S} = 295.8 \quad s_S^2 = 6865.78 \quad SD = 82.86$$

The standard error of the sample mean difference, $\bar{d} = \bar{F} - \bar{S}$, is $SE(\bar{d}) = \sqrt{2s_p^2/n}$ where now s_p^2 denotes the *pooled* sample variance, $s_p^2 = (s_F^2 + s_S^2)/2 = (3559.32 + 6865.78)/2 = 5212.84$, hence $s_p = 72.21$. The final step in the analysis is again to calculate a 95% confidence interval for the population mean difference as

$$\bar{d} \pm 2SE(\bar{d}) = (-11.3, 102.0).$$

The numerical value of $\bar{d}$ is the same as in the paired analysis, but the standard error is not. As a result, the confidence interval is centred on the same value, 45.4, but it is substantially wider than before. Under the parallel group scenario, the result of the trial would have been inconclusive, because the confidence interval spans zero.

In general, there is no necessity for the two sample sizes to be the same in a parallel group trial. Table 6.3 sets out the calculations in the general case of two unequal sample sizes, n_x and n_y, say. Although the notation and formulae are now somewhat inelegant, you should be able to satisfy

Table 6.3. Stages in the calculation of a 95% confidence interval for the population mean difference between two separate batches of data with sample sizes n_x and n_y.

Data	$x_1, x_2, \ldots, x_{n_x}$
Sample mean	$\bar{x} = n_x^{-1} \sum_{i=1}^{n_x} x_i$
Sample variance	$s_x^2 = (n_x - 1)^{-1} \sum_{i=1}^{n_x} (x_i - \bar{x})^2$
Data	$y_1, y_2, \ldots, y_{n_y}$
Sample mean	$\bar{y} = n_y^{-1} \sum_{i=1}^{n_y} y_i$
Sample variance	$s^2 = (n_y - 1)^{-1} \sum_{i=1}^{n_y} (y_i - \bar{y})^2$
Pooled variance	$s_p^2 = \{(n_x - 1)s_x^2 + (n_y - 1)s_y^2\}/(n_x + n_y - 2)$
Mean difference	$\bar{d} = \bar{x} - \bar{y}$
Standard error	$SE(\bar{d}) = \sqrt{s_p^2 \times (n_x^{-1} + n_y^{-1})}$
95% confidence interval	$\bar{d} \pm 2 \times SE(\bar{d})$

6.4 Crossover designs

yourself that if $n_x = n_y = n$, they reduce to the expressions given above for the asthma data. And the earlier health warning still applies.

As noted above, it is legitimate to consider the asthma data as a paired design because the order of presentation of the two drugs was randomized. However, because asthmatic symptoms can be exacerbated by unfavourable weather conditions, it may have turned out that there were systematic differences in response between the two time periods in which the drugs were administered and the PEF measured. The paired analysis, although legitimate, would then have been inefficient, because the difference between the two time periods would have inflated the standard error of the estimated treatment effect.

A crossover trial is one in which the order of presentation of two (or more) treatments is considered as a factor in the experiment, and included in the analysis accordingly. To re-analyse the asthma data as a crossover trial, we need to use more information than Table 6.1. The first seven rows are results from children who received Formoterol in the first time period and Salbutamol in the second time period, whilst in the remaining six rows the order of presentation was reversed.

In Section 6.3.1 we quoted a point estimate $\bar{D} = 45.4$ for the mean difference between PEF values obtained using Formoterol and Salbutamol, respectively. To take account of any effect of time period on the outcome, we need first to compute two such estimates. From the first seven rows of Table 6.1 we get $\bar{D}_1 = 30.7$, and for the last six rows we get $\bar{D}_2 = 62.5$. What are these actually estimating? Let δ denote the population mean difference between PEF values recorded under drugs F and S administered in the same time period: this is the parameter of scientific interest. Also, let τ denote the population mean difference between PEF recorded under the same drug, but administered in the first and second time period. Then:

$$\bar{D}_1 \text{ is estimating } \delta + \tau,$$
$$\bar{D}_2 \text{ is estimating } \delta - \tau.$$

This suggests considering the following pair of equations,

$$\hat{\delta} + \hat{\tau} = 30.7,$$
$$\hat{\delta} - \hat{\tau} = 62.5.$$

You can easily check that the solution to this pair of equations is

$$\hat{\delta} = 46.6, \quad \hat{\tau} = -15.9.$$

So the crossover analysis gives a slightly different answer than does the paired analysis. Is it significantly different? And has the crossover analysis produced a useful gain in precision? We will give the formal statistical answer to these questions in Chapter 7, as an example of statistical modelling.

Senn (2002) gives a good discussion of the design and analysis of crossover trials and their use in clinical research.

6.5 Comparing more than two treatments

Both the parallel group and paired designs generalize immediately to experiments whose aim is to compare more than two treatments.

The generalization of the parallel group design, usually called the completely randomized design, allocates the experimental units randomly amongst the $k > 2$ treatments. Specifically, if n_j denotes the number of units to be allocated to treatment j, and $n = n_1 + \cdots + n_k$, then we first pick a random sample of n_1 out of the n units to receive treatment 1, then a random sample of n_2 out of the remaining $n - n_1$ to receive treatment 2, and so on.

The generalized version of the paired design is called the randomized block design. It requires the experimenter to identify blocks of k experimental units and allocate the treatments $1, 2, \ldots, k$ randomly amongst the k units in each block.

Calculation templates analogous to those set out in Tables 6.2 and 6.3 can be written down, but are not especially illuminating. Rather than pursue this topic here, we shall revisit it in Section 7.6.5, where we will show how the analysis of an experiment to compare more than two treatments can be cast as a special case of statistical modelling.

7
Statistical modelling: the effect of trace pollutants on plant growth

7.1 Pollution and plant growth

When the water supply used to irrigate food crops contains traces of contaminant, there is a danger that the growth of the crop will be affected adversely. The likelihood of this happening is greater in the case of intensive farming, where irrigation and the application of pesticides or weedkillers may go hand in hand, and run-off water from irrigated crops may be recycled to reduce waste.

Figure 7.1 shows the results of an experiment conducted in the CSIRO's Centre for Irrigation Research, Griffith, New South Wales, in which young safflower plants were grown in water deliberately contaminated with specified concentrations of glyphosate, a widely used weedkiller. The inputs to each run of the experiment were the concentration of glyphosate in parts per million (ppm) and whether the glyphosate had been added to distilled or to tap water. Using distilled water, six runs were made with no added glyphosate and three at each of the concentrations $0.053, 0.106, 0.211, 0.423, 0.845, 1.690$ and 3.380 parts per million. The experiment was then repeated using tap water instead of distilled water. The output from each run was the total root length of a batch of 15 plants. In more statistical terminology, we are dealing with $n = 2 \times (6 + 3 \times 7) = 54$ *replicates* of an experiment with a single, continuously varying *response* (total root length), a continuously varying design variable (glyphosate concentration) and a binary *factor* (type of water).

An obvious qualitative conclusion from these data is that total root length tends to decline as the concentration of glyphosate increases. Our main aim in building a *model* for the data is to quantify this relationship, whilst a secondary aim is to establish whether the relationship is affected by which type of water is used. Notice that the non-zero concentrations used in the experiment represent successive dilutions of the maximum concentration, by a factor of two in each case. This is a practically convenient design, but we shall see later that it is also a sensible choice from the point

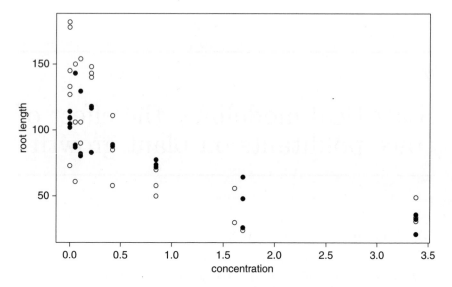

Fig. 7.1. Total root length of batches of safflower plants grown in water contaminated with specified concentrations of glyphosate (parts per million). Solid dots are for plants grown in distilled water, open circles are for plants grown in tap water.

of view of modelling the relationship between glyphosate concentration and root length.

7.2 Scientific laws

Scientific laws are expressions of quantitative relationships between variables in nature that have been validated by a combination of observational and experimental evidence.

As with laws in everyday life, accepted scientific laws can be challenged over time as new evidence is acquired. The philosopher Karl Popper summarizes this by emphasizing that science progresses not by proving things, but by disproving them (Popper, 1959, p. 31). To put this another way, a scientific hypothesis must, at least in principle, be falsifiable by experiment (iron is more dense than water), whereas a personal belief need not be (Charlie Parker was a better saxophonist than John Coltrane).

7.3 Turning a scientific theory into a statistical model: mechanistic and empirical models

We now revisit our discussion in Chapter 2 of the lab experiment designed to illustrate one of Newton's laws of motion, namely that *a body in free fall under the influence of the Earth's gravity experiences a constant*

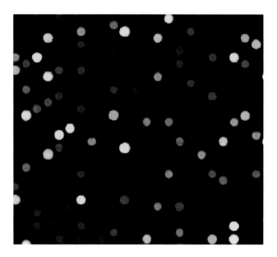

Fig. 4.1. A microarray image. Each spot on the array represents a gene. The colour of the spot codes for the ratio of expression levels of that gene in response to two different stimuli, from bright green at one extreme to bright red at the other.

Fig. 5.1. Part of the Rothamsted research station, showing field trials in progress.

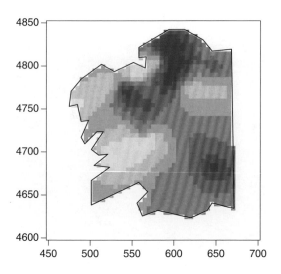

Fig. 10.14. Mean predicted pollution surface for the Galicia lead pollution data. The colour-scale runs from 2.0 (red) through yellow to 10.0 (white). Mapped values range from 2.8 to 7.8.

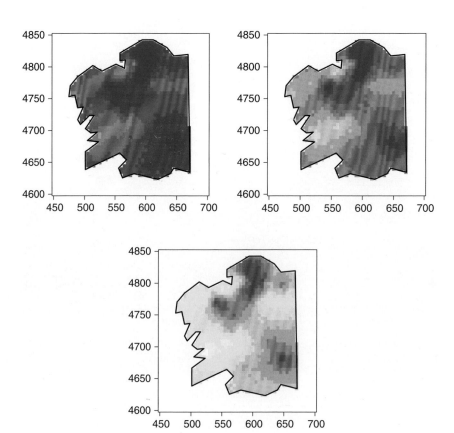

Fig. 10.15. Lower quartile, median and upper quartile predicted pollution surfaces for the Galicia lead pollution data. The colour-scale runs from 2.0 (red) through yellow to 10.0 (white).

acceleration. To turn this *scientific law* into a *mathematical model* we use the integral calculus to deduce that the relationship between the vertical distance, d, fallen by a body initially at rest and the time, t, since the body began falling can be expressed by the equation

$$d = \frac{1}{2}gt^2. \tag{7.1}$$

If we so choose, we can equally well rearrange equation (7.1) as in Chapter 2 to give

$$y = \beta x, \tag{7.2}$$

where now y denotes time, $x = \sqrt{d}$ and $\beta = \sqrt{2/g}$. As we discussed in Chapter 2, to a mathematician, (7.1) and (7.2) are saying precisely the same thing. But (7.2) is more useful because it shows us how we can easily falsify Newton's law (if, of course, it is indeed false) by picking three different values of x, recording the corresponding values of y and plotting the three points (x, y) on a graph to see whether they do or do not lie along a straight line. But this is a counsel of perfection. Can we measure distance and time precisely? Can we set the necessary equipment up in a vacuum, so as to eliminate completely the extraneous effects of air resistance? The most we can expect is that the points lie *approximately* on a straight line. But this begs the question of what we mean by 'approximately'. Now the value of the re-expression (7.2) becomes apparent: x is an input to the experiment, and if we so choose we can repeat the experiment holding the value of x fixed; in contrast, y is an output from the experiment and repetitions of the experiment would not generate identical values of y because of the various imperfections in our equipment, our experimental technique and so forth. The two panels of Figure 7.2 show two (fictitious) datasets which could be obtained by running an experiment once under each of two different experimental conditions and twice under a third. The pair of results from the repeated experiment give us an idea of how large is the experimental error. In the left-hand panel, the results would probably lead us to conclude that these data are compatible with an underlying linear relationship between the input x and the output y. In the right-hand panel, three of the four results are the same as in the left-hand panel, but the repeat run now suggests that the experimental error is extremely small and we might be more inclined to conclude that the underlying relationship is non-linear. Of course, in reality such a small experiment would not convince a sceptical audience one way or the other, but the important point is that, at least in principle, if we can repeat an experiment under controlled conditions we can quantify the size of the experimental error and thereby distinguish between models which are compatible with the data and models which are not. This is (half of – see below!) the essence

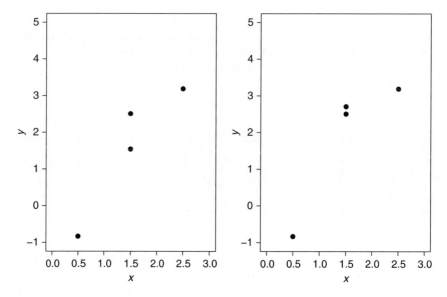

Fig. 7.2. Two fictitious datasets: in the left-hand panel the data are compatible with an underlying linear relationship between the input x and the output y; in the right-hand panel they are not.

of the statistical method, and echoes Popper's philosophy in the sense that we may be prepared to rule out many incompatible models (i.e., falsify the theories that led to them). But demonstrating that a model is compatible with the data is not to say that the theory behind it is true: there may be other theories that lead to different, but equally compatible, models and they can't all be true.

Why 'half of'? The answer is that statisticians use the mathematics of probability to quantify experimental error, and as a result are rarely in a position to declare that a given model cannot possibly have generated the data at hand, only that it would be very unlikely to have done so, in a sense that can be quantified using probability theory (as discussed in Chapter 3).

A statistical model typically includes both deterministic (systematic) and stochastic (random) components. Sometimes, as is the case for the lab experiment discussed in Chapter 2, the deterministic component can be justified by appeal to well-understood scientific laws. The same can apply, but much more rarely in practice, to the stochastic component. In our physics lab experiment, we recognized a role for stochastic variation to represent the unpredictability of human reaction time, but we had no scientific reason to choose a particular probability distribution to describe this effect. So we settled for a simple assumption, justified to some extent by looking at the data, that the variation in reaction time could be modelled as a random variable that behaved in the same way whatever the value of x.

Models derived from scientific laws are often called *mechanistic* models, to distinguish them from *empirical models*, whose only claim is that they are compatible with the data. Mechanistic models are generally preferable to empirical models, but are not always available; in their absence, empirical models can still contribute to scientific progress, and have an honourable track record of doing so for at least the past hundred years.

7.4 The simple linear model

Suppose that we want to investigate the relationship between an input variable x and an output variable y. The simplest possible mathematical model for the relationship between x and y is a straight line, or *linear relationship*,

$$y = \alpha + \beta x, \tag{7.3}$$

where α represents the *intercept* and β the *slope* of the line (see Figure 7.3). This model admits three qualitatively different possibilities. The first is that there is no relationship between x and y, in which case $\beta = 0$. The second is that as x increases, so does y, in which case β is positive. The third is that as x increases, y decreases, in which case β is negative.

Now suppose that we conduct n independent replicates of the experiment. In the ith replicate we set the value of the input variable to be x_i, and observe the resulting value, y_i, of the output variable. To establish whether the data are compatible with model (7.3) we need only plot the points $(x_i, y_i) : i = 1, \ldots, n$ and see whether they do or do not lie along a straight line (assuming that there are at least three different numerical values amongst the x_i). But if the output variable is subject to experimental error, this test is too stringent.

To allow for experimental error, or more generally for stochastic variation in y over experimental replicates with the value of x held fixed, we extend the *mathematical* model (7.3) to the *statistical* model

$$Y_i = \alpha + \beta x_i + Z_i : i = 1, \ldots, n. \tag{7.4}$$

The differences between (7.3) and (7.4) are subtle but important. Most obviously, the additional term Z_i on the right-hand side of (7.4) measures the amount by which each experimental result (x_i, Y_i) deviates from the straight line $Y = \alpha + \beta x$. Recall from Chapter 2 that we use upper-case letters to indicate stochastic variables in the model, and lower-case letters to indicate deterministic variables. The values of x_i are treated as deterministic because the experimenter can fix them in advance.

Recall also that this notational convention allows us to distinguish between the *model* (7.4), and the *data*, $(x_i, y_i) : i = 1, \ldots, n$, which result from it. The *lower-case* y_i represents the actual value of the output variable

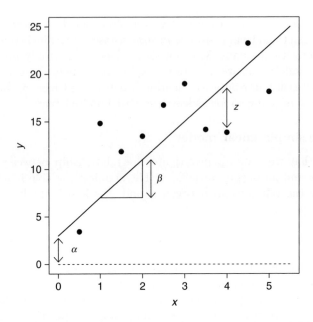

Fig. 7.3. Schematic representation of the simple linear model (7.4).

obtained from the ith replicate, whereas the *upper-case* Y_i represents the stochastic process that generates the observed value y_i each time we run the experiment. Finally, the number of replicates is made explicit to emphasize that, in contrast to (7.3), our ability to identify appropriate values for α and β in (7.4) depends on how large n is. Figure 7.3 indicates the different elements of (7.4) schematically.

To complete the specification of (7.4), we need to make some assumptions about the nature of the stochastic variation in the Z_i. The standard assumptions are:

A1: each Z_i has mean zero;

A2: the Z_i are independent;

A3: each Z_i has standard deviation σ, irrespective of the value of x_i;

A4: each Z_i is Normally distributed.

These assumptions are invoked tacitly by almost any piece of software that claims to calculate a 'line of best fit', but there is no guarantee that they will hold good in any particular application; we return to this point in Section 7.7.1 below. For the moment, we note only that they are numbered in decreasing order of importance. Only A1 is essential, because it implies that the relationship between the explanatory variable x and the mean of the response Y is indeed linear. A2 and A3 are important because they

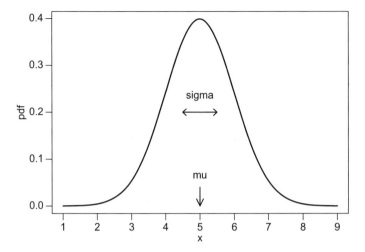

Fig. 7.4. The Normal probability density function with mean μ and standard deviation σ.

imply that the standard errors for the intercept and slope parameters α and β given by standard software are reliable. A4 is only important if the focus of scientific interest lies in the detailed behaviour of the deviations from the relationship between x and the mean of Y, rather than in the relationship itself. Nevertheless, practical experience has shown that A4 does often hold, at least to a good approximation. The so-called *Normal* probability distribution (note the upper case N – there is nothing abnormal about other probability distributions) is described by a symmetric, bell-shaped probability density function, an example of which is shown in Figure 7.4. This function includes two parameters, the *mean*, conventionally denoted by the Greek letter μ, and the *standard deviation*, σ. When this distribution is used to describe the stochastic variation about a modelled value, for example the Z_i in equation (7.4), the mean is automatically set to zero. Approximately 68% of the total probability lies within plus or minus one standard deviation of the mean, and approximately 95% within plus or minus two standard deviations. The alert reader may notice, correctly, that this last statement echoes our earlier definition of a 95% confidence interval as a sample mean plus or minus two standard errors.

Figure 7.5 shows three simulated datasets, each generated by the same model, namely (7.4) with $\alpha = 3$, $\beta = 2$, $n = 10$, $x_i = 1, 2, \ldots, 10$ and $\sigma = 0.5$. The data $(x_i, y_i) : i = 1, \ldots, 10$ shown in the three panels of Figure 7.5 differ only because the corresponding sets of realized values $z_i : i = 1, \ldots, 10$ differ. Note that each panel included two straight lines. The solid line is the underlying 'true' relationship, $y = 3 + 2x$, and is identical in all three panels. The dashed lines are the so-called lines of best

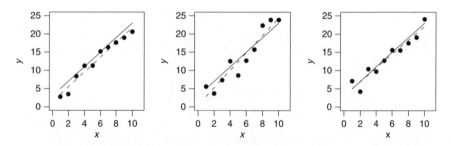

Fig. 7.5. Three simulated datasets. The underlying model is the same in all cases (see text for details). In each panel the solid dots show the data $(x_i, y_i) : i = 1, \ldots, 10$, the solid line is the underlying linear relationship $y = 3 + 2x$, and the dashed line is the line of best fit.

fit, and differ across the three panels. These are estimates of the underlying 'true' relationship, and they differ across the three panels because they are calculated from the data, $(x_i, y_i) : i = 1, \ldots, 10$ and inherit the stochastic character of the y_i.

7.5 Fitting the simple linear model

Fitting the simple linear model to a set of data $(x_i, y_i) : i = 1, \ldots, n$ means choosing numerical values of α and β to give the 'line of best fit' to the data. Because these numerical values are only estimates of the 'true' line, we write them as $\hat{\alpha}$ and $\hat{\beta}$. When the random variation in the data is small, meaning that in assumption A3 above, σ is negligible compared to the deterministic variation in the x_i, fitting by eye does the job. Otherwise, and in any event if we want an automatic, reproducible procedure, we need a criterion to measure how 'far' the data points (x_i, y_i) are from the fitted line $y = \hat{\alpha} + \hat{\beta}x$.

An objective approach to defining an appropriate criterion is to apply the method of maximum likelihood, as described in Chapter 3, to the statistical model defined by assumptions A1 to A4 above. This results in the rule: estimate $\hat{\alpha}$ and $\hat{\beta}$ to minimize the quantity

$$LS(\alpha, \beta) = \sum_{i=1}^{n}(y_i - \alpha - \beta x_i)^2. \tag{7.5}$$

In equation (7.5), each term on the right-hand side is just the squared value of z_i, the implied value of the z-value shown on Figure 7.3. This seems a sensible criterion, but is certainly not the only one. If we replaced A1 to A4 by a different set of assumptions, the likelihood principle would lead us to a different criterion, as it should.

Note in particular that (7.5) measures how 'far' a point (x_i, y_i) is from the line $y = \alpha + \beta x$ by the *vertical* distance between the two. The justification for doing so is that the model is intended to describe how the outcome y can vary for any fixed value of x; hence, if our model is correct, the points do not lie exactly on the line because of variation in the y-direction. When both x and y are randomly varying output variables, then minimizing (7.5) is arguably not the right thing to do.

The 'LS' in equation (7.5) stands for 'least (sum of) squared (residuals)', and the method of choosing α and β to minimize $LS(\alpha, \beta)$ is called *least squares* estimation. The likelihood principle tells us that this is the best possible strategy when assumptions A1 to A4 all hold.

7.6 Extending the simple linear model

The simple linear model often fits experimental data surprisingly well. However, there is certainly no guarantee that it will do so in any particular application. In this section, we discuss some ways in which the model can be made more flexible.

7.6.1 *Transformations*

In the gravity experiment described in Chapter 2, we fitted a simple linear model to the data by defining the x-variable to be the *square root* of the vertical distance fallen by the ball-bearing. This is an example of a *transformation* of the input variable. Transformations of this kind are unexceptionable; if we can measure and fix the value of x, then we can equally measure and fix the value of any function, or transformation, of x. The square-root transformation was suggested by the form of Newton's law as $d = \frac{1}{2}gt^2$, or equivalently, $y = bx$, where $y = t$, $x = \sqrt{d}$ and $b = 2/g$ is a constant. Superficially, we could equally well have re-expressed Newton's law as $y = cx$ where now $y = t^2$, $x = d$ and $c = 2/g$. However, had we done so, there would have been no justification for simply adding a random term to the right-hand side to explain why the observed data did not exactly follow the theoretical linear relationship between x and y: instead, the effects of random variation in human reaction time would have entered into the model in a very complicated way as random additions to the square root of y. The message to take from this is that any decision to transform the output variable needs careful consideration.

Notwithstanding the conclusion of the previous paragraph, if our data arise in a context where there is no well-accepted scientific law to describe the relationship between the input and output variables, choosing a transformation of either or both variables empirically is often a useful way to make the simple linear model a better approximation to the underlying scientific truth, whatever that may be.

The glyphosate data illustrate this idea. The left-hand panel of Figure 7.6 is identical to Figure 7.1 except for the aspect ratio of the

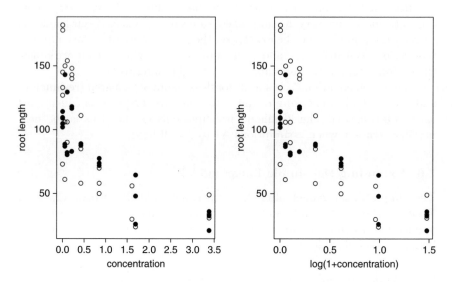

Fig. 7.6. Total root length of batches of safflower plants grown in water contaminated with specified concentrations of glyphosate. In the right-hand panel, the x-axis has been changed from concentration to $\log(1 + \text{concentration})$. Solid dots are for plants grown in distilled water, open circles are for plants grown in tap water.

diagram, whereas the right-hand panel shows the same data, but now using the x-axis to represent $\log(1 + \text{concentration})$. Why did we choose this particular transformation? Firstly, the experimental protocol of using successive two-fold dilutions suggested that the experimenter had in mind that the effect of varying concentration might be multiplicative rather than additive; secondly, the $+1$ preserves the zero point, since $\log(1) = 0$. The transformation has reduced the strength of the visual impression of curvature in the relationship, but probably has not eliminated it altogether. Also, the outcome variable, total root length, appears to vary more at low concentrations than at high concentrations, in violation of assumption A3. There are several ways to deal with this second concern. One is to ignore it; the standard least-squares fitting procedure will still give sensible estimates of α and β. A better way is to modify the procedure to recognize that the variability in Y depends on the value of the input variable, x.

One way to modify the procedure is to replace the least squares criterion (7.4) by a *weighted least squares* criterion,

$$WLS(\alpha, \beta) = \sum_{i=1}^{n} (y_i - \alpha - \beta x_i)^2 / v(x_i). \qquad (7.6)$$

Ideally, the $v(x_i)$ in (7.6) should be proportional to the variance of the output variable Y when the input variable takes the value x_i, this being another application of the method of maximum likelihood.

Another possibility is to explore the effects of possible transformations of the output variable. This has two effects: it changes the shape of the curve (the aim being to find a transformation such that the curve becomes a straight line); and it changes the way in which the variability in Y does or does not depend on the value of x (the aim being to achieve a constant variance). There is no guarantee that these two aims are mutually compatible. Nevertheless, and again primarily for illustration, Figure 7.7 shows the effect of transforming both the x-variable, to $\log(1 + \text{concentration})$ as before, and the y-variable, to $\log(\text{root length})$.

Purely from an empirical point of view, the right-hand panel of Figure 7.7 looks reasonably well described by the simple linear model. Whether or not it gives a biologically satisfying explanation of the way in which trace contamination with glyphosate leads to deterioration in plant growth is a much harder question, and can only be answered through a dialogue between statistician and scientist. Statistical modelling is straightforward when the data are from a tightly controlled experiment operating according to well-accepted scientific laws. In other situations it is less straightforward, and arguably as much an art as a science.

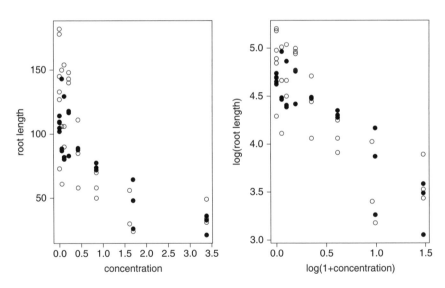

Fig. 7.7. Total root length of batches of safflower plants grown in water contaminated with specified concentrations of glyphosate. In the right-hand panel, the x-axis has been changed from concentration to $\log(1 + \text{concentration})$ and the y-axis from root length to $\log(\text{root length})$. Solid dots are for plants grown in distilled water, open circles are for plants grown in tap water.

7.6.2 *More than one explanatory variable*

So far, we have only discussed models for the relationship between the response variable and a single explanatory variable. If we have two explanatory variables, say x and u, we can easily extend the simple linear model (7.4) to a model of the form

$$Y_i = \alpha + \beta x_i + \gamma u_i + Z_i : i = 1, \ldots, n.$$

To visualize this model, the analogue of Figure 7.3 would be a collection of points (x, u, Y) distributed around an inclined plane in three-dimensional space, rather than around a straight line in two-dimensional space.

If we have more than two explanatory variables, we obtain the *general linear model*,

$$Y_i = \alpha + \beta_1 x_{1i} + \beta_2 x_{2i} + \cdots + \beta_p x_{pi} + Z_i : i = 1, \ldots, n. \qquad (7.7)$$

A direct visualization is not available, but the principle is the same. We can rewrite (7.7) as a pair of equations,

$$Y_i = \mu(x_{1i}, \ldots, x_{pi}) + Z_i : i = 1, \ldots, n,$$
$$\mu(x_{1i}, \ldots, x_{pi}) = \alpha + \beta_1 x_{1i} + \cdots + \beta_p x_{pi}. \qquad (7.8)$$

Then, $\mu(x_{1i}, \ldots, x_{pi})$ represents the model's best prediction of the value of a future response Y when the explanatory variables take the values $x_{1i}, \ldots, x_{pi}$, whilst Z_i, called the *residual*, is the difference between an actual response Y_i and its predicted value $\mu(x_{1i}, \ldots, x_{pi})$ according to the model. As we shall see in Section 7.7.1 this representation of the model leads to a way of checking the compatibility of the model for any number of explanatory variables.

7.6.3 *Explanatory variables and factors*

Recall from earlier chapters that the terms *explanatory variables* and *factors* are both used to describe variables that affect the outcome of an experiment and that, conventionally, the former is used when the variable in question takes values over a continuous range (e.g., the concentration of glyphosate) and the latter when it takes only a discrete set of values (e.g., distilled or tap water). Both kinds can be accommodated within the general linear model (7.7).

7.6.4 *Reanalysis of the asthma trial data*

The way to include factors within the general linear model (7.7) can most easily be explained through a specific example, such as the asthma trial that we discussed in Chapter 6. In that example, the response variable Y_i was the PEF measured on the ith child and our objective was to compare the mean values of PEF obtained under two different drug treatments, Formoterol (F) and Salbutamol (S). In Chapter 6 we represented this by writing the two mean values as μ_F and μ_S. But since we were mainly interested in the difference, if any, between the effectiveness of the two drugs it would have been at least as natural to write the two means as $\mu_F = \alpha$ and $\mu_S = \alpha + \delta$, where δ represents the population mean difference in PEF achieved under administration of F and S. Now, if we define a variable x to take the value $x = 0$ for a PEF measurement taken after administration of S and $x = 1$ for a measurement taken after administration of F, it follows that both μ_F and μ_S can be defined by the single equation $\mu_i = \alpha + \delta \times x_i : i = 1, \ldots, 26$, and the responses Y_i can be modelled as

$$Y_i = \alpha + \delta x_i + Z_i : i = 1, \ldots, 26, \tag{7.9}$$

in other words as a special case of the general linear model.

Because the asthma trial used a paired design, we cannot assume that the pairs of residuals Z_i from the same child are independent. But if we subtract each child's S measurement from their F measurement to obtain a single value, D_i say, for each child, then equation (7.9) reduces to

$$D_i = \delta + Z_i^* : i = 1, \ldots, 13, \tag{7.10}$$

where now the Z_i^* are independent. Equation (7.10) is the basis of the simple analysis that we described in Section 6.3.1 and can now be seen as a very simple example of the general linear model as defined by equation (7.7).

We shall now revisit the crossover analysis of these same data. In Section 6.4, we defined δ as above, and τ as the population mean difference in PEF in time periods 1 and 2. Now, in addition to the explanatory variable x that indicates which drug was administered ($x = 0/1$ for S and F, respectively), we define a second explanatory variable $u = 0$ for a PEF measurement taken in the second time period and $u = 1$ for a PEF measurement taken in the first time period. This gives the extended model

$$Y_i = \alpha + \delta x_i + \tau u_i + Z_i : i = 1, \ldots, 26. \tag{7.11}$$

If we now take the difference between the two measured PEF values on each of the 13 children, again subtracting their S measurement from their F measurement, we eliminate α as before, but not τ, because the difference between the corresponding values of u will be either plus or minus one, depending on whether the child in question received F in the first or the second time period, respectively. So the model for the difference D_i is now

$$D_i = \delta + \tau v_i + Z_i^* : i = 1, \ldots, 13, \tag{7.12}$$

where the explanatory variable v takes values plus or minus one. This is again a special case of the general linear model (7.7).

We can now use the model (7.12) to answer the questions that we left hanging at the end of Section 6.4.

Recall that the paired analysis of the asthma data gave the point estimate $\hat{\delta} = 45.4$ with associated 95% confidence interval $(22.9, 67.9)$. In Section 6.4 we gave what seemed an intuitively reasonable point estimate of δ, adjusting for a time-period effect, as $\hat{\delta} = 46.6$. The difference between 45.4 and 46.6 is not clinically significant, but is the estimate from the crossover analysis any more precise? Part of the output from fitting the model (7.12) in R is

```
Coefficients:
            Estimate Std. Error
(Intercept)   46.61     10.78
v            -15.89     10.78

Residual standard error: 38.74 on 11 degrees of freedom
```

The rows labelled (Intercept) and v give the estimates of the parameters δ and τ, respectively. The results confirm that the intuitive estimates $\hat{\delta} = 46.6$ and $\hat{\tau} = -15.9$ quoted in Section 6.4 are in fact the maximum likelihood estimates under the assumed model. More importantly, the 95% confidence interval for δ is now $46.61 \pm 2 \times 10.78 = (25.0, 68.2)$, to one decimal place. The width of this confidence interval is 43.2, only slightly narrower than the width 45.0 obtained from the paired analysis. Notice, however, that the 95% confidence interval for the time-period effect, τ is $(-37.4, 5.7)$, which includes zero. So in this trial, there is no compelling evidence of a time-period effect, and it is therefore not unreasonable that adjusting for it through the crossover analysis gives only a modest improvement in precision.

7.6.5 Comparing more than two treatments

As a second example, we pick up the brief discussion of Section 6.5 concerning the analysis of comparative experiments involving more than two treatments. Let p denote the number of treatments, and define p explanatory variables $x_1, \ldots, x_p$ amongst which x_k takes the value $x_{ik} = 1$ if the ith subject is allocated to treatment k, $x_{ik} = 0$ otherwise. Then, the response $Y_i : i = 1, \ldots, n$ can be modelled as

$$Y_i = \beta_1 x_{i1} + \ldots + \beta_p x_{ip} + Z_i, \tag{7.13}$$

again a special case of the general linear model (7.7).

As a specific example, we consider the 12 measured expression levels of Gene 2 in Table 4.1. As the response variable y, we use log-base-two-transformed expression level. For a first analysis we codify the four combinations of strain (wild-type or Ca resistant) and calcium challenge (low or high) as a single factor with four levels:

1 wild-type   low
2 wild-type   high
3 Ca resistant low
4 Ca resistant high

There is more than one way to translate this into a linear model of the form (7.13). Perhaps the most obvious is to define four explanatory variables, say x_1, x_2, x_3 and x_4, as indicator variables for the four treatments, i.e., $x_1 = 1$ when the allocated treatment is 1, zero otherwise, and similarly for x_2, x_3 and x_4. Then, the corresponding parameters $\beta_1, \ldots, \beta_4$ in (7.13) represent the mean values of the response under each of the four treatment allocations. Part of the output from fitting this model in R is as follows:

```
Coefficients:
           Estimate Std. Error
treatment1  2.6269    0.1636
treatment2  2.5702    0.1636
treatment3  2.7448    0.1636
treatment4  2.9433    0.1636

Residual standard error: 0.2834 on 8 degrees of freedom
```

The column headed Estimate gives the estimated values $\hat{\beta}_1$ to $\hat{\beta}_4$, whilst the column headed Std.Error gives their estimated standard errors; these all take the same value because the software uses a pooled estimate of the variances under all four treatments, and the replication is the same for all treatments. The Residual standard error is slightly misnamed. It is the pooled estimate of the standard deviation, i.e., $s^2_{pooled} = 0.2834^2 = 0.0803$, and the reference to 8 degrees of freedom is simply saying that four regression parameters have been estimated from 12 observations, i.e., $8 = 12 - 4$.

Often, treatment means themselves are less interesting than differences between treatment means. If there is a natural baseline treatment relative to which the other treatments are to be compared, we can translate this into a linear model by defining $x_1 = 1$ whatever the allocated treatment, $x_2 = 1$ for treatment 2 and zero otherwise, $x_3 = 1$ for treatment 3 and zero otherwise, $x_4 = 1$ for treatment 4 and zero otherwise. With these definitions, the software output becomes:

```
              Estimate  Std. Error
(Intercept)    2.6269    0.1636
treatment2    -0.0567    0.2314
treatment3     0.1178    0.2314
treatment4     0.3163    0.2315
```

Residual standard error: 0.2834 on 8 degrees of freedom

Now, the row labelled (Intercept) relates to the parameter β_1, which still represents the estimated mean response under treatment 1, whereas the remaining three rows relate to parameters β_2, β_3 and β_4, which now represent differences between the mean response under treatments 2 and 1, 3 and 1, 4 and 1, respectively. The values of $\hat{\beta}_1$ and s^2_{pooled} are the same in both cases, as they should be, because all we are doing is fitting two different representations of the *same* model, i.e., one in which the four treatment means are different. Notice that the one case in which the standard error is smaller than the corresponding parameter estimate is for the difference in mean response between treatments 4 and 1, which compares the Ca-resistant strain under a high calcium challenge with the wild-type strain under a low calcium challenge. The 95% confidence interval for this difference is $0.3163 \pm 2 \times 0.2315 = (-0.1467, 0.7793)$. Recall that the aim of the original experiment was to find genes that are involved in regulating the plant's response to calcium: specifically, an 'interesting' gene is one that shows a relatively large differential response between low and high calcium challenges in the Ca-resistant strain and a relatively low differential response in the wild-type strain. Hence, although the confidence interval is inconclusive, it at least hints that this particular gene may be so involved.

We can do a better job of formulating our model to capture the scientific aim by incorporating the factorial structure of the four treatments. This leads to the following definitions of the four explanatory variables:

x_1 1

x_2 1 for Ca-resistant strain, 0 otherwise

x_3 1 for high calcium challenge, 0 otherwise

x_4 1 for Ca-resistant strain under high calcium challenge, 0 otherwise

The corresponding interpretation of the β-parameters, using the terminology from Section 5.4, is:

β_1 mean response for wild-type strain under low calcium challenge

β_2 main effect of strain (Ca-resistant vs. wild-type strain)

β_3 main effect of calcium challenge (high vs. low)

β_4 interaction

The corresponding software output is:

```
                  Estimate  Std. Error
(Intercept)        2.62694     0.16361
strain             0.11782     0.23138
challenge         -0.05675     0.23138
strain:challenge   0.25525     0.32722
```

`Residual standard error: 0.2834 on 8 degrees of freedom`

Note in particular that the interpretation of the estimated interaction effect (`strain:challenge`) is that the Ca-resistant strain shows a greater differential response from the low to the high challenge than does the wild-type strain, by an estimated amount 0.25525 with associated approximate 95% confidence interval $0.25525 \pm 2 \times 0.32722 = (-0.39919, 0.90969)$. The result is therefore inconclusive. This is unsurprising. For the purposes of this illustrative analysis, the four genes whose results are given in Table 4.1 were selected arbitrarily. The number of genes involved in regulating the calcium challenge is likely to be small, making it very unlikely that we would find such a gene in a random sample of four from the 22,810 possible candidates.

7.6.6 What do these examples tell us?

An important conclusion to be drawn from the analyses described in Sections 7.6.4 and 7.6.5 is that the same software can be used to fit models that include both explanatory variables and factors. This in turn implies that there is no need to learn cumbersome algebraic formulae to analyse different experimental designs.

7.6.7 Likelihood-based estimation and testing

The method of maximum likelihood can be used to estimate the parameters in any statistical model. The likelihood function can also be used to compare models, and in particular to decide whether an explanatory variable warrants inclusion in the model by using a likelihood ratio test as described in Section 3.4. Suppose that we have identified a provisional model with p explanatory variables $x_1, \ldots, x_p$ and associated regression parameters $\beta_1, \ldots, \beta_p$ and want to decide whether another explanatory variable, x_{p+1} say, should be added to the model. Then, the provisional model becomes a special case of the extended model, with the constraint that $\beta_{p+1} = 0$. The discussion of likelihood ratio testing in Section 3.4 now applies, with $\theta = (\beta_1, \ldots, \beta_p, \beta_{p+1})$ and $\theta_0 = (\beta_1, \ldots, \beta_p, 0)$. Hence, if we write L_0 and L_1 for the maximized values of the log-likelihood associated with the provisional and extended models, respectively, we would reject the provisional model in favour of the extended model, i.e., include the explanatory variable x_{p+1}, if D is bigger than the value in Table 3.2 corresponding to $m = 1$ and the chosen level of significance (conventionally, 5%).

Sometimes we need to consider adding a set of explanatory variables, rather than a single one. Suppose, for example, that the provisional model is a simple linear regression, whilst the extended model allows the slope parameter, β, to depend on a factor with three levels. Then, the extended model is

$$Y = \alpha + \beta_j x + Z : j = 1, 2, 3,$$

and the provisional model assumes that $\beta_1 = \beta_2 = \beta_3$. Equivalently, we can parameterize the extended model as $\theta = (\beta_1, \beta_2 - \beta_1, \beta_3 - \beta_1)$, in which case $\theta_0 = (\beta_1, 0, 0)$. The key point is that whatever parameterization we choose to use, the extended model has two more parameters than the provisional model, and the appropriate critical value for the likelihood ratio test can be found in Table 3.2 in the column corresponding to $m = 2$.

7.6.8 Fitting a model to the glyphosate data

Based on the discussion of the glyphosate data in Section 7.6, we use as our provisional model a linear regression of $Y = \log(\text{root length})$ on $x = \log(1 + \text{concentration})$ with separate intercepts for distilled and tap water, but a common slope, hence

$$Y = \alpha_1 + \alpha_2 t + \beta x + Z, \tag{7.14}$$

where $t = 0$ for plants grown in distilled water and $t = 1$ for plants grown in tap water, hence α_2 denotes the difference between the mean of Y for plants grown in tap water rather than in distilled water, irrespective of the concentration of glyphosate. Parameter estimates and standard errors are given in the following table:

Parameter	Estimate	Standard error
α_1	4.7471	0.0639
α_2	0.0609	0.0771
β	-0.9456	0.0795

Note that the 95% confidence interval for α_2 is $0.0609 \pm 2 \times 0.0771$, or $(-0.0933, 0.2151)$. This interval includes zero, suggesting that the difference between tap and distilled water is not statistically significant. Probably of more interest is the estimate of β, which describes how the mean root length declines with increasing concentration of glyphosate in the water. The 95% confidence interval for β is $-0.9456 \pm 2 \times 0.0795$, or $(-1.1046, -0.7866)$, which clearly excludes zero. To complete the picture, the 95% confidence interval for α_1 is $(4.6193, 4.8749)$.

The maximized log-likelihood for this model is $L_0 = 69.67$. One way to extend the model is to allow both the intercept and slope to depend on whether the plants are grown in distilled or tap water. This replaces the single parameter β by two parameters, β_1 and β_2. The maximized log-likelihood for the extended model is $L_1 = 69.88$, hence the likelihood ratio statistic to compare the provisional and extended models is $D = 2 \times (69.88 - 69.67) = 0.42$. Since this is comfortably less than 3.84, we have no reason to reject the provisional model. We therefore conclude that the relationship between root length and glyphosate concentration, as described by the parameter β, does not depend on whether the glyphosate is added to distilled or to tap water.

Purely on statistical grounds, we could equally argue that the provisional model (7.14) could be simplified by setting $\alpha_2 = 0$. We choose not to do so because the hypothesis that mean root length is the same for plants grown in distilled and in tap water is not of any scientific interest, nor is it particularly plausible as tap water contains small quantities of a range of nutrient minerals that may well assist growth. In contrast, it is both plausible and of some interest to ask whether the deleterious effect of trace concentrations of glyphosate does or does not depend on the chemical composition of the uncontaminated water supply.

To assess the precision of our estimate of the parameter of interest, β, the estimate and its confidence interval are difficult to interpret because they relate to transformed values of both the response and the explanatory variable. A more useful way to express the result obtained from the model (7.14) is to see how the fitted response curves change as β varies over its confidence interval. Writing g and r for glyphosate concentration and mean root length, respectively, our fitted model for plants grown in distilled water is $\log r = \alpha_1 + \beta \log(1 + g)$, or equivalently

$$r = \exp\{\alpha_1 + \beta \log(1 + g)\}. \tag{7.15}$$

For plants grown in tap water, we would need to replace α_1 in (7.15) by $\alpha_1 + \alpha_2$. Figure 7.8 shows four versions of (7.15), corresponding to all four combinations of α and β at the lower and upper ends of their respective confidence intervals, together with the data from plants grown in distilled water, and gives a more easily interpretable summary of how precisely we have been able to describe the relationship between root length and glyphosate concentration.

7.7 Checking assumptions

Models rest on assumptions that may or may not be compatible with the data. All statistical models acknowledge that the features of scientific

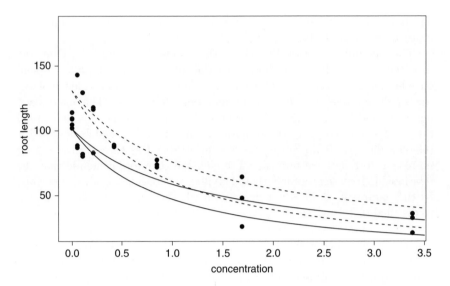

Fig. 7.8. Total root length of batches of safflower plants grown in distilled water contaminated with specified concentrations of glyphosate (solid dots), and fitted mean response curves with intercept and slope parameters set at each end of their respective confidence intervals. Solid lines and dashed lines correspond to the lower and upper confidence limits, respectively, for α. Within each line style, the lower and upper curves correspond to the lower and upper confidence limits, respectively, for β.

interest in the results of an experiment may be partially obscured by unpredictable variation in the outcome. We represent this by including in our model one or more random variables. A symbolic mnemonic for this is

$$\text{DATA} = \text{SIGNAL} + \text{NOISE}$$

This applies literally in the case of the general linear model (7.8), in which the Y_i are the data, the $\mu(x_{ii}, \ldots, x_{pi})$ are the signal and the Z_i are the noise. Although we are unable to observe the noise directly, we are able to estimate it by substituting into the model our best estimate of the signal, and deducing the corresponding values of the noise. For the general linear model, we do this by calculating the *fitted values*,

$$\hat{\mu}_i = \hat{\alpha} + \hat{\beta}_1 x_{1i} + \cdots + \hat{\beta}_p z_{pi} \qquad (7.16)$$

and the *residuals*,

$$\hat{z}_i = y_i - \hat{\mu}_i. \qquad (7.17)$$

CHECKING ASSUMPTIONS

Diagnostic checking is the process by which we compare the data with the fitted model in ways that are designed to reveal any major incompatibilities. It turns out that many of the most effective ways of doing this involve analysing the residuals in various ways, as we now describe. We continue to work within the specific context of the general linear model, although the ideas apply more generally.

7.7.1 Residual diagnostics

Recall the four assumptions on which the general linear model rests:

A1: each Z_i has mean zero;

A2: the Z_i are independent;

A3: each Z_i has standard deviation σ, irrespective of the value of x_i;

A4: each Z_i is Normally distributed.

Collectively, these assumptions require that the residuals should behave in every respect as if they were an independent random sample from a Normal distribution with zero mean and constant variance.

Figure 7.9 shows a simple but instructive example, devised by the statistician Frank Anscombe, of how the pattern of the residuals can reveal a range of departures from the assumptions A1 to A4. In all four cases,

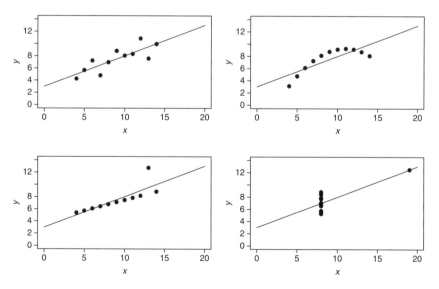

Fig. 7.9. Anscombe's quartet: four synthetic datasets that give the same fitted linear regression model.

fitting a simple linear regression model to the data gives the following table of results:

Parameter	Estimate	Standard error
α	3.0001	1.1247
β	0.5001	0.1179

However, the four datasets give very different patterns of residuals, here shown in Figure 7.10 as scatterplots of the residuals against their corresponding fitted values. In the top-left panel, the residuals behave as we would wish them to, i.e., an apparently random pattern, suggesting a good fit between model and data. In the top-right panel, there is a very clear pattern, whose explanation is that we have fitted a linear regression when the true relationship between x and y is non-linear. The bottom-left panel suggests a perfect linear relationship, but with one aberrant data point, which in practice might have arisen either from a failure of the experiment or a simple mis-recording of the result. Finally, the bottom-right panel gives no evidence for or against the model, since with only two distinct values of x in the data, one of which is unreplicated, we can neither check the assumed linearity of the relationship (A1) nor the constancy of the variance (A3). An experimental design in which half of the measurements are made at opposite extremes of the relevant range of x is actually the most efficient design possible *if* the modelling assumptions are known to hold in advance of the experiment, but is a very risky design otherwise.

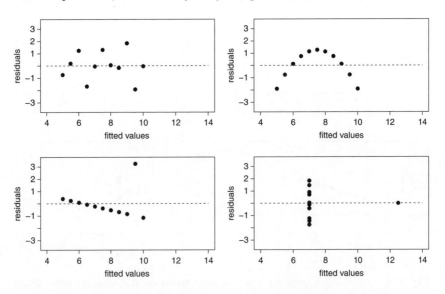

Fig. 7.10. Anscombe's quartet: residual against fitted value plots.

Once you have seen Figure 7.9, the subsequent residual plot, Figure 7.10, holds no surprises. However, if the model includes more than one explanatory variable, the analogue of Figure 7.9 requires a multidimensional representation that is more difficult to interpret with two explanatory variables, and well nigh impossible with more than two, whereas the analogue of Figure 7.10 is always two-dimensional.

A scatterplot of residuals against fitted values is the single most useful kind of diagnostic plot. However, it does not directly address assumption A3: independence of residuals. One common circumstance in which A3 is violated is when repeated measurements are made on the same, or related experimental units. When the number of experimental units is small, for example when data from several different labs are to be combined for analysis, independence can usually be restored by including labs as a factor in the model. Otherwise, as for example in human studies involving all members of a large number of families, non-routine statistical methods are usually needed to deal with the resulting dependence between measurements made on either the same person, or on different members of the same family.

The independence assumption can also be violated when data are obtained in time order and there is any form of carry-over between successive runs of the experiment, for example because of a drift over time in the micro-environment in which the experiment is conducted. For this reason it is *always* a good idea to plot residuals against the time at which the corresponding datum was obtained whenever this information is available.

An extreme form of time-ordered data is a single time series, an example of which is the Bailrigg temperature data that we discussed briefly in Section 4.5.1. The dominant feature of that series is the seasonal variation in temperature. As we shall see in Chapter 9, a good model for the seasonal pattern is a general linear model using sine and cosine terms as explanatory variables. Here, we simply show in Figure 7.11 the residual time series obtained by subtracting the fitted seasonal curve from the data. Notice that at various points along the series there are relatively long runs of positive residuals followed by runs of negative residuals. This suggests a lack of independence, although we postpone confirmation of this until Chapter 9.

7.7.2 *Checking the model for the root-length data*

Our fitted model for the glyphosate data is given by equation (7.14) with parameter values $\hat{\alpha}_1 = 4.7471$, $\hat{\alpha}_2 = 0.0609$ and $\hat{\beta} = -0.9456$. Using these values, we can convert the 54 data points y_i to fitted values,

$$f_i = \hat{\alpha}_1 + \hat{\alpha}_2 t_i + \hat{\beta} x_i$$

and residuals, $r_i = y_i - f_i$.

Figure 7.12 shows a scatterplot of residuals against fitted values, using different plotting symbols to distinguish between plants grown in distilled

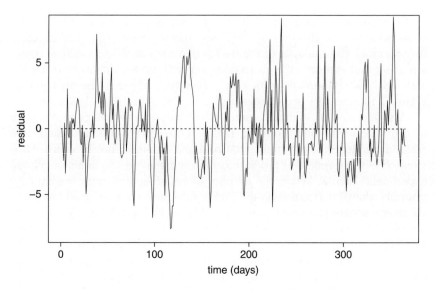

Fig. 7.11. Time series of residuals, after subtracting a seasonal curve from the Bailrigg temperature data.

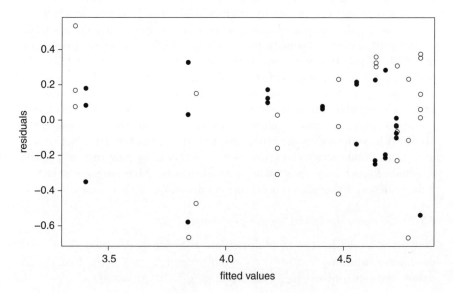

Fig. 7.12. Residuals against fitted values plot for model (7.14) fitted to the glyphosate data. Residuals corresponding to plants grown in distilled and in tap water are shown by solid dots and open circles, respectively.

7.8 An exponential growth model

We have seen through our analysis of the root-length data how we can use transformations of either explanatory variables or the response variable to convert a non-linear relationship into a linear one. Another example, widely encountered in the natural sciences, is the phenomenon of *exponential growth*. To make this tangible, we consider a set of data consisting of five-yearly estimates of the total population of the world, between 1950 and 2010. Call these y_t, where t runs from one (year 1950) to 13 (year 2010). Table 7.1 shows the values of y_t, together with successive differences, $d_t = y_t - y_{t-1}$, and ratios, $r_t = y_t/y_{t-1}$.

The differences vary substantially, by a factor of approximately two within the time span of the data. The ratios show much smaller relative changes, although they are somewhat lower towards the end of the sequence. Exponential growth is a mathematical model of the form

$$\mu_t = \exp(\alpha + \beta t), \qquad (7.18)$$

where, as in the pair of equations (7.8), μ_t denotes the mean of Y_t. In this model, $\exp(\alpha)$ is the *initial size* of the population, i.e., the value of Y_t at time $t = 0$, and β is the *growth rate*, expressed as a proportion; hence, for

Table 7.1. Estimated world population, in thousands, at five-yearly intervals between 1950 and 2010 (source: United Nations Population Network).

Year	Population	Difference	Ratio
1950	2529346	NA	NA
1955	2763453	234107	1.093
1960	3023358	259905	1.094
1965	3331670	308312	1.102
1970	3685777	354107	1.106
1975	4061317	375540	1.102
1980	4437609	376292	1.093
1985	4846247	408638	1.092
1990	5290452	444205	1.092
1995	5713073	422621	1.080
2000	6115367	402294	1.070
2005	6512276	396909	1.065
2010	6908688	396412	1.061

example, if the population grows by 5% per year, $\beta = 0.05$ (strictly, the growth rate is $\exp(\beta) - 1$ but this is approximately equal to β if β is a small fraction). Now, equation (7.18) implies that

$$\log \mu_t = \alpha + \beta t. \tag{7.19}$$

This suggests that taking $\log Y_t$ as the response, we could use a *linear* model to estimate the growth rate, β, and the initial true population size, $\exp(\alpha)$. If we do this using the least squares criterion (7.5), we are implicitly assuming that errors in the transformed responses $\log Y_t$ are additive, i.e., that $\log Y_t = \log \mu_t + Z_t$. This may or may not be a reasonable assumption, but we can at least check whether or not it is compatible with the data using the methods described in Section 7.7.1.

To see how this works for the world population data, Figure 7.13 plots the logarithms of population estimates against year. The linear model turns out not to be a good fit, as a consequence of a slowing in the growth rate towards the end of the time-period covered by the data. To emphasize this, Figure 7.13 includes the result of fitting a linear model to the first nine observations only, covering the period 1950 to 1990. This gives an almost perfect fit, but extrapolates poorly beyond 1990.

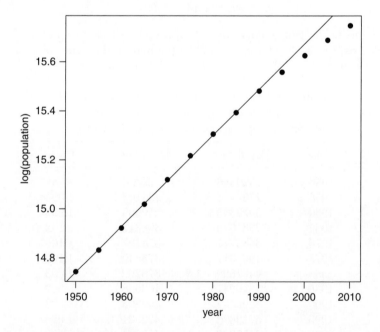

Fig. 7.13. Estimates of log-transformed total world population, at five-yearly intervals from 1950 to 2010. The line corresponds to the fit of an exponential growth model to the estimates up to 1990 only.

AN EXPONENTIAL GROWTH MODEL

With a little ingenuity, we can accommodate this behaviour as yet another special case of the general linear model (7.7). Let x denote the years, subtracting 1950 from each so that the time period begins at zero. Hence, $x = 0, 5, 10, \ldots, 60$. Let y denote the corresponding log-transformed population estimates. Now define a second explanatory variable u to take the values

$$u = 0 \quad 0 \quad 0 \quad 0 \quad 0 \quad 0 \quad 0 \quad 0 \quad 5 \quad 10 \quad 15 \quad 20$$

If we now fit the model

$$y = \alpha + \beta x + \gamma u + Z, \tag{7.20}$$

the regression parameters have the following interpretation. The intercept, α, represents the logarithm of population size in 1950 (recall that the data are only estimates); β represents the annual relative growth rate of the population between 1950 and 1990; finally, γ represents the increase or decrease in the relative growth rate between 1990 and 2010. An extract from the output obtained when fitting model (7.20) using R, in the now-familiar form, is:

```
Coefficients:
            Estimate  Std. Error
(Intercept) 14.7406651 0.0032456
x            0.0187380 0.0001285
u           -0.0056059 0.0003670

Residual standard error: 0.005373 on 10 degrees of freedom
```

We therefore estimate the 1950 population size as $\exp(14.7406651) = 2522258$ thousand, close but not identical to the value 2529346 recorded in Table 7.1. Which is the better estimate is a moot point; the value 2522258 is likely to be the more accurate if the assumed model (7.20) is correct, and conversely. The model-based estimate does, however, carry with it an estimate of its precision, again assuming the model to be correct. To calculate a confidence interval for the required quantity, $\exp(\alpha)$, we simply calculate the corresponding confidence interval for α and exponentiate both ends of the interval. Using our standard rule, an approximate 95% confidence interval for α is

$$\hat{\alpha} \pm 2 \times SE(\hat{\alpha}) = 14.7406651 \pm 2 \times 0.0032456 = (14.73417, 14.74716),$$

hence an approximate 95% confidence interval for $\exp(\alpha)$ is

$$((\exp(14.73417), \exp(14.74716)) = (2505929, 2538693),$$

which comfortably includes the tabulated value 2529346.

Our estimate and approximate 95% confidence interval for the annual growth rate between 1950 and 1990 are

$$\hat{\beta} \pm 2 \times SE(\hat{\beta}) = 0.0187380 \pm 2 \times 0.0001285 = (0.018481, 0.018995),$$

i.e., a fraction under 2% per year.

Finally, the change in the annual growth rate since 1990 is estimated to be

$$\hat{\gamma} \pm 2 \times SE(\hat{\gamma}) = -0.0056059 \pm 2 \times 0.0003670 = (-0.0063399, -0.0048719),$$

a decrease in the growth rate of between 0.5% and 0.6%, i.e., from a little under 2% to a little under 1.5%. The ouptut shown above does not allow us to calculate a confidence interval for the post-1990 annual growth rate, $\beta + \gamma$, but this can be extracted from the full output to give an approximate 95% confidence interval $(0.01257637, 0.01368784)$, i.e., we estimate the post-1990 annual growth rate to be between about 1.3% and 1.4%.

The fit of this 'split line' model to the data is shown graphically in Figure 7.14. All of the individual data points lie on or very close to the fitted split line.

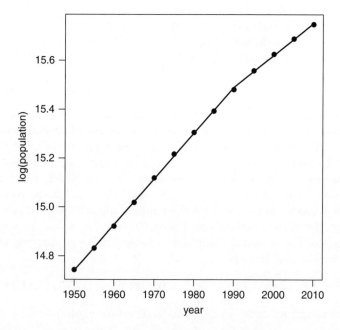

Fig. 7.14. Estimates of log-transformed total world population, at five-yearly intervals from 1950 to 2010. The line corresponds to the fit of an exponential growth model that allows a change in the growth rate at 1990.

7.9 Non-linear models

Using transformations to convert non-linear relationships to linear ones is not always possible. As an example, consider how a biological population, for example, of reproducing bacteria, might grow when there is a limit on the amount of food available to sustain the growing population. If the initial population size is sufficiently small that food is abundant, the exponential growth model might hold in the early stages of growth, but as food becomes a limiting factor, the growth rate might slow down. One model that captures this behaviour is

$$\mu_t = \gamma \exp(\alpha + \beta t)/\{1 + \exp(\alpha + \beta t)\}. \tag{7.21}$$

This model still describes approximately exponential growth in the early stages, but the growth rate slows as the population approaches an upper limit, γ. Figure 7.15 illustrates this.

The so-called *logistic* growth model (7.21) often gives a good description of the growth of natural populations, but cannot be transformed into a linear model of the form (7.8). Models of this kind are called *intrinsically non-linear*.

Intrinsically non-linear models violate the first assumption, A1, of the general linear model. They may still satisfy assumptions A2 to A4, in which case the method of maximum likelihood still gives the least squares criterion, here

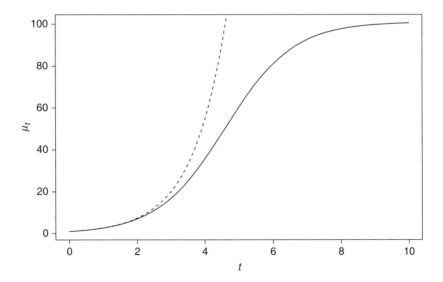

Fig. 7.15. A comparison between exponential (dashed line) and logistic (solid line) growth models.

$$LS(\alpha, \beta, \gamma) = \sum_t \{Y_t - \mu_t(\alpha, \beta, \gamma)\}^2,$$

as the optimal method for estimating α, β and γ.

7.10 Generalized linear models

In the previous section, we showed how the exponential growth model could be converted into a linear model by transforming the expression for the mean response, from $\mu_t = \exp(\alpha + \beta t)$, to $\log \mu_t = \alpha + \beta t$. However, we also warned that this gave no guarantee that the corresponding transformation of the response, from Y_t to $\log Y_t$, would result in a model that would satisfy the remaining assumptions A2 to A4 of the general linear model.

Generalized linear models pick up on this concern by unlinking assumption A1, which concerns the mean response, from assumptions A3 and A4, which concern the random variation in the response about its mean. A full discussion of generalized linear models goes beyond the scope of this book, but their flavour can be conveyed by two examples, each of which deals with a situation in which the representation of a response as a mean plus a Normally distributed residual is clearly, and grossly, wrong.

7.10.1 *The logistic model for binary data*

Our first example concerns experiments in which the response, Y, is binary, i.e., $Y = 0$ or 1. For example, in toxicology experiments we might vary the dose, x, of a toxin and observe whether this does ($Y = 1$) or does not ($Y = 0$) result in the death of the animal to which the dose was applied. Now, the mean of Y, $\mu(x)$ say, represents the probability that the animal will be killed, and must lie between 0 and 1 whatever the value of x. A linear model, $\mu(x) = \alpha + \beta x$, does not satisfy this constraint. However, if we transform $\mu(x)$ to the quantity

$$\eta(x) = \log[\mu(x)/\{1 - \mu(x)\}], \qquad (7.22)$$

then $\eta(x)$ can take any value at all: when $\mu(x)$ is close to zero $\eta(x)$ is large and negative; when $\mu(x)$ is close to one $\eta(x)$ is large and positive. So a linear model for $\eta(x)$ might not be unreasonable. Putting $\eta(x) = \alpha + \beta x$ on the left-hand side of (7.22) and solving for $\mu(x)$ gives

$$\mu(x) = \exp(\alpha + \beta x)/\{1 + \exp(\alpha + \beta x)\}. \qquad (7.23)$$

Figure 7.16 gives examples of the function $\mu(x)$ defined by equation (7.23). The value of $\mu(x)$ necessarily lies between 0 and 1 for all values of x. The effect of increasing or decreasing α is to shift the curve to the left or right, respectively, whilst the effect of increasing or decreasing β is to make the characteristic S-shape of the curve steeper or shallower, respectively.

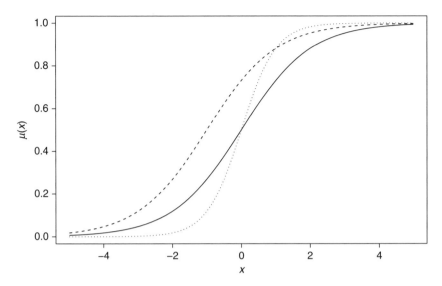

Fig. 7.16. Examples of the inverse logit function (7.23). Parameter values for the three curves are $\alpha = 0, \beta = 1$ (solid line), $\alpha = 1, \beta = 1$ (dashed line) and $\alpha = 0, \beta = 2$ (dotted line).

The transformation defined by equation (7.22) is called the *logit* or *log-odds* transformation, whilst equation (7.23) defines the *inverse logit*. The inverse logit transformation has the effect of constraining the mean of Y to lie between 0 and 1, as it must, and the only sensible way to model random variation in Y is as a series of trials with α, β and x held fixed, in which the proportion of trials that result in $Y = 1$ is given by equation (7.23). The resulting model is called the *logistic-linear regression model*, often abbreviated to *logistic model*, for a binary response. As with its linear counterpart, we can extend the logistic model by including more than one explanatory variable on the right-hand side of (7.23), or by transforming the explanatory variables. For example, Figure 7.17 shows the functions

$$\mu(x) = \exp(-5 + 2x)/\{1 + \exp(-5 + 2x)\}$$

and

$$\mu(x) = \exp(-5 + x + 0.01x^4)/\{1 + \exp(-5 + x + 0.01x^4)\}$$

over values of x between 0 and 10. Notice how by including x^4 as well as x in the model, we obtain an asymmetric rather than a symmetric S-shaped curve.

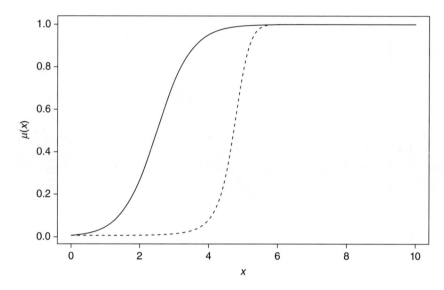

Fig. 7.17. Linear (solid line) and quartic (dashed line) inverse logit functions. Note the asymmetry in the quartic case.

7.10.2 *The log-linear model for count data*

Our second example is a model for experiments whose response Y is a non-negative count. For example, we might measure the number of seizures experienced by an epileptic patient over a fixed follow-up period after administration of a dose x of a particular medication. If the counts are large, we might be willing to assume a linear model for the relationship between Y and x, but if the counts are small a linear model may predict a mean count less than zero, which is clearly impossible. One response to this is to transform Y to $\log Y$, but this runs into problems because $\log 0$ is undefined. A pragmatic solution is to transform to $\log(1+Y)$, which transforms zero to zero as previously illustrated in our analysis of the glyphosate data. A more principled approach is to transform the mean, $\mu(x)$, to $\eta(x) = \log \mu(x)$ and assume a linear model for $\eta(x)$, i.e., $\eta(x) = \alpha + \beta x$. The inverse transformation is

$$\mu(x) = \exp(\alpha + \beta x). \tag{7.24}$$

To model the random variation in Y, the same argument against a model of the form $Y = \mu(x) + Z$ applies as in the case of binary Y, albeit with less force if $\mu(x)$ is large. Amongst the many options available the simplest, and the usual starting point, is to assume that Y follows a *Poisson* probability distribution. This specifies the probability of observing any non-negative integer value of Y as a function of its mean, μ. The algebraic form of the Poisson distribution is

$$p(y) = \exp(-\mu)\mu^y/y! : y = 0, 1, \ldots, \qquad (7.25)$$

where $y!$ (read as 'y factorial') is a shorthand notation for the quantity $y \times (y-1) \times \cdots \times 1$, with the convention that $0! = 1$. Figure 7.18 shows several examples of the Poisson distribution as superimposed frequency polygons. The distribution is positively skewed, but the skewness decreases as μ increases.

The class of models in which Y follows a Poisson distribution whose mean depends on one or more explanatory variables x according to equation (7.24), or its obvious extension to the case of multiple explanatory variables, is called the *Poisson log-linear regression model*.

7.10.3 *Fitting generalized linear models*

In formulating a linear model for a particular set of data we need to choose which of the available explanatory variables should be included in the model, and whether each of the selected variables should be included in its original or a transformed form (or, indeed, both, as for example in a quadratic model, $Y = \alpha + \beta x + \gamma x^2$).

In a *generalized linear model*, we need to make the same choice *and* two others. Firstly, the chosen combination of parameters and explanatory variables does not define the mean response itself but rather a transformation of the mean response, called the *linear predictor* and conventionally denoted

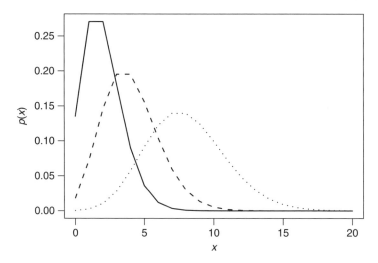

Fig. 7.18. Poisson probability distributions with mean $\mu = 2$ (solid line), $\mu = 4$ (dashed line) and $\mu = 8$ (dotted line).

$\eta(x)$. Thus, for example, the logistic model uses $\eta = \log\{\mu/(1-\mu)\}$, whilst the log-linear model uses $\eta = \log\mu(x)$. The transformation from μ to η is called the *link function*.

The other choice to be made is what probability distribution we should use to describe the random variation in Y. For binary Y, there *is* only one such distribution, since μ gives the probability that $Y = 1$ and the probability that $Y = 0$ must therefore be $1 - \mu$. For a count Y, one choice is the Poisson distribution, but there are many others. The chosen distribution is called the *error distribution*.

Generalized linear models have transformed the practice of applied statistics since they were introduced by Nelder and Wedderburn (1972). One reason for this is that their theoretical development was followed rapidly by their incorporation into a dedicated statistical software package, GLIM (Generalized Interactive Linear Modelling), and later into a wide variety of more general statistical packages. This was possible because Nelder and Wedderburn showed that essentially the same algorithm could be used to compute maximum likelihood estimates of the parameters in *any* generalized linear model.

7.11 The statistical modelling cycle: formulate, fit, check, reformulate

Statistical modelling is useful when, for whatever reason, the statistical analysis protocol for a scientific experiment or observational study has not been fully specified before the study has been conducted. This is typically the case in what might loosely be called discovery science, in which the investigator anticipates finding relationships amongst variables that will be measured during the course of a study, but cannot anticipate precisely what form those relationships might take.

In contrast, modelling is generally considered inappropriate for the initial, or *primary* analysis of data from tightly controlled experiments such as an agricultural field trial or a clinical trial whose purpose is to provide an unequivocal answer to a specific question, typically of the form: do experimental treatments give significantly different average responses? However, even in this context, modelling may have a role to play in follow-up, or *secondary* analyses.

The distinction between primary and secondary analysis is explicit in clinical trial protocols. Consider, for example, a completely randomized trial to compare a novel with a standard therapy, and suppose that the response from each subject in the trial is a single measure, y say, of their improvement in health following treatment. The primary analysis of a trial of this kind would almost certainly be a two-sample t-test, as described in Section 6.3.2, with the intention of giving a robust answer to the specific question: is the average value of y significantly higher for patients receiving the novel therapy than for patients receiving the standard therapy? As discussed in Section 5.5, this protocol merits the description 'robust' because the random allocation of subjects to treatments automatically delivers a statistically valid test. A conclusion that the novel therapy gives significantly the better average result may, if other conditions are satisfied, result in the novel therapy replacing the current standard. However, it immediately invites a number of supplementary questions: is the novel therapy superior for all patients? is it more or less effective in particular subgroups? in the case of a chronic condition, is its effectiveness long-lasting or transient? These and other questions can be addressed in secondary analyses, and modelling can help to provide answers.

After a study has been conducted and the data collected, the process of statistical modelling includes three distinct phases of activity. The first stage is to formulate a model for the data that meets two criteria: it is capable of providing an answer to the scientist's question; and it is not self-evidently incompatible with the data. Meeting the second of these involves exploratory data analysis to decide which of the available explanatory variables might be worth including in the model, whether their relationship with the response variable can be captured by a linear model, and whether there are features of the pattern of variation in the data that require special consideration; for example, whether apparent outliers represent coding errors or genuinely aberrant behaviour that, according to context, might or might not be of particular interest.

The second stage is to fit the model, i.e., to estimate its parameters and to confirm, or not as the case may be, that variables provisionally included in the model in stage 1 should be retained. The best advice for this second stage is to use likelihood-based, rather than ad hoc, methods whenever possible. Most reputable software packages use likelihood-based methods.

The third stage is diagnostic checking, using the residuals from the fitted model. In an ideal world, this will lead to the happy conclusion that the model is compatible with the data, i.e., is a 'good fit'. In practice, diagnostic checking may reveal inconsistencies between model and data that were not apparent earlier, in which case the model should be reformulated and the modelling cycle repeated. It cannot be overemphasized that likelihood-based methods of estimation and testing are optimal *when the underlying modelling assumptions are correct* but they do not address goodness-of-fit.

Many of the models discussed in this chapter fall within the scope of the generalized linear model. The authoritative book on this topic is McCullagh and Nelder (1989). Dobson (2001) is a more accessible introduction.

8
Survival analysis: living with kidney failure

8.1 Kidney failure

The kidney performs the vital function of filtering and thereby removing toxins that would otherwise accumulate in the body. Kidney failure is therefore a life-threatening condition. For most forms of kidney failure, the most effective treatment is a kidney transplant. A successful transplant provides approximately 50% of normal kidney function, which is more than sufficient to restore the patient to good health. However, for a number of reasons including a chronic shortage of donor organs, not every patient can receive a transplant and others may have to wait for several years before a suitable donor organ becomes available. For these patients, an alternative treatment is *dialysis*. This consists of using an artificial device to filter toxins, either from the blood (haemo-dialysis) or from the peritoneal fluid (peritoneal dialysis).

In peritoneal dialysis, several litres of a sugar solution are introduced into the peritoneal cavity through a tube surgically implanted in the patient's abdomen. The sugar solution gradually extracts toxins through a process of osmosis across the peritoneal membrane. After some hours, the solution is removed and is replaced by fresh solution. In one version of the treatment (continuous ambulatory peritoneal dialysis, CAPD) the removal and replacement of fluid is done by the patient, typically four times a day, using a simple drain-and-fill sequence driven by the force of gravity. In a second version (automated peritoneal dialysis, APD) the drain-and-fill operation is effected by an electrical pump programmed to run overnight.

Table 8.1 is an extract from a clinical database maintained over a number of years by Dr Peter Drew, a renal physician at the Maelor Hospital, Wrexham, North Wales. The data relate to patients being treated by peritoneal dialysis. For each patient, the data give the time, in days, from initiation of treatment to the patient's death. How can we decide which, if either, version of peritoneal dialysis has the better survival prognosis?

Table 8.1. Data on survival of kidney failure patients receiving peritoneal dialysis. Columns 1 and 4 give the time (in days) for which the patient is known to have survived. Columns 2 and 5 indicate whether the patient died at the time indicated (coded 1), or was last observed alive and may or may not have died subsequently (coded 0). Columns 3 and 6 give each patient's age, in years, at the time the dialysis was started.

APD			CAPD		
Time (days)	Dead	Age (years)	Time (days)	Dead	Age (years)
3444	0	41	147	1	55
3499	0	35	422	1	45
6230	0	41	5096	0	46
1324	1	67	3353	0	37
6230	0	29	5415	0	30
709	1	54	551	1	63
6230	0	42	4851	0	44
6230	0	36	4452	0	25
391	1	74	3927	0	52
2790	1	70	4760	0	23

8.2 Estimating a survival curve

At first glance, the way to answer the question just posed is obvious – just use the methods described in Chapter 6 for analysing data from simple comparative experiments. But the caption to Table 8.1 explains why this is not so straightforward. The outcome of interest is the time until a defined event, death after starting dialysis, takes place. But for some patients, the event has not yet happened.

Data of this kind are called *time-to-event* or *survival* data. Their characteristic feature is that the event of interest is not always observed exactly, but may be *censored*, meaning that we know only that the survival time is greater than a given value. In other settings, we may know only that the survival time lies between two given values or, more rarely, that it is smaller than a given value. The three forms of censoring are called right-censored, interval-censored and left-censored, respectively. We will only discuss the analysis of right-censored data, and use the term censored without qualification to mean right-censored.

The key to analysing censored data is to ask a subtly different question. Rather than comparing different experimental treatments on the basis of the average survival times of subjects in each treatment group, we estimate for each treatment and each possible survival time, t say, the probability that a subject will survive for at least a time t. If t is zero, this probability automatically takes the value 1. As t increases, the survival probability decreases. In many applications the survival probability must eventually fall to zero, but this is not always so. For example, if the event of interest were the age at which a person suffers acute kidney failure, some people would never experience the event.

A plot of the estimated survival probability against survival time is called a *survival curve*. In the absence of censoring, a natural estimate is, for each t, the proportion of survivors, i.e., subjects who have not experienced the event of interest by time t. To illustrate, consider the following fictitious set of uncensored survival times:

$$0.9 \quad 1.3 \quad 1.5 \quad 3.5 \quad 4.9$$

The five survival times have been ordered from smallest to largest. So, for any value of t less than 0.9, the proportion of survivors is 1, for t between 0.9 and 1.3 the proportion of survivors is $4/5 = 0.8$, for t between 1.3 and 1.5 the proportion is $3/5 = 0.6$, and so on. Figure 8.1 shows this estimate, with the five data points as circles. The survival curve, $S(t)$ say, starts at $S(0) = 1$ and is piece-wise constant, decreasing by $1/5 = 0.2$ at each observed survival time.

Now consider a superficially perverse route to the same answer. How does a subject survive until time 1.4? First they have to survive until time 0.9. From the data, we estimate this probability to be $p_1 = 4/5 = 0.8$. Now, having survived until time 0.9, they also have to survive until time 1.4. But of the four remaining subjects at time 0.9, three survive until time 1.4. So we estimate this second probability to be $p_2 = 3/4 = 0.75$.

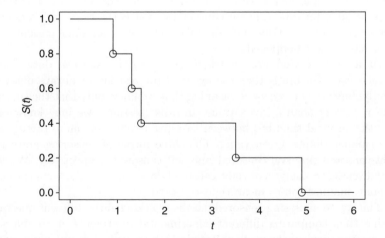

Fig. 8.1. The estimated survival curve from five uncensored survival times.

In the formal language of probability theory, p_1 is an estimate of the unconditional probability of survival until time 0.9, and p_2 is an estimate of the conditional probability of survival until time 1.3, given survival until time 0.9. To estimate the unconditional probability of survival until time 1.4, we need to multiply the two estimates, to give

$$p_1 \times p_2 = \frac{4}{5} \times \frac{3}{4} = \frac{3}{5} = 0.6.$$

Now, consider the same set of numbers as before, except that the second and fourth survival times are censored at 1.3 and 3.5, which we indicate as

$$0.9 \quad 1.3^+ \quad 1.5 \quad 3.5^+ \quad 4.9$$

Now we can't simply estimate survival probabilities by observed proportions because, for example, we don't *know* what proportion survived until time 1.4 – it could be 3/5 or 4/5. But we can use a simple modification of our perverse method. With the new data, we still estimate the probability of survival until time 0.9 as $p_1 = 4/5 = 0.8$, because we do still know that the second survival time is bigger than 0.9. But our estimate of the conditional probability of survival until time 1.4 given survival until time 0.9 is now 3/3, not 3/4, because we can no longer include the censored observation in our calculation. So our estimate of the unconditional probability of survival until time 1.4 becomes

$$p_1 \times p_2 = \frac{4}{5} \times \frac{3}{3} = \frac{12}{15} = 0.8.$$

This feels a bit like cheating. It's as if we are pretending that the survival time censored at 1.3 is actually at least 1.4. But as we continue in this way, the calculation adjusts itself automatically in a sensible way. For example, the estimate of the unconditional probability of survival until time 1.6 is

$$p_1 \times p_2 \times p_3 = \frac{4}{5} \times \frac{3}{3} \times \frac{2}{3} = 0.533$$

because we know that 2 out of 3 patients alive at time 1.3 survived at least until time 1.6. Quite properly, this estimate lies between the values 0.4 and 0.6 that would hold if the censored survival time 1.3 were known to be less than 1.6 or greater than 1.6, respectively.

Table 8.2 shows the complete calculation of the survival curve, allowing for both censored survival times, whilst Figure 8.2 shows the resulting estimate $S(t)$ as a function of survival time, t.

This method of estimating a survival curve is called the Kaplan–Meier estimate, after the two statisticians who introduced it, Richard Kaplan and Paul Meier (Kaplan and Meier, 1958).

We can use the same method to estimate two survival curves from Dr Drew's data, one for each of the two versions of dialysis. The complete

Table 8.2. Calculating the Kaplan–Meier estimate of a survival curve.

Range of t	Value of survival curve
0.0 to 0.9	$\frac{5}{5} = 1$
0.9 to 1.3	$\frac{5}{5} \times \frac{4}{5} = 0.8$
1.3 to 1.5	$\frac{5}{5} \times \frac{4}{5} \times \frac{3}{3} = 0.8$
1.5 to 3.5	$\frac{5}{5} \times \frac{4}{5} \times \frac{3}{3} \times \frac{2}{3} = 0.533$
3.5 to 4.9	$\frac{5}{5} \times \frac{4}{5} \times \frac{3}{3} \times \frac{2}{3} \times \frac{2}{2} = 0.533$
4.9 or more	$\frac{5}{5} \times \frac{4}{5} \times \frac{3}{3} \times \frac{2}{3} \times \frac{2}{2} \times \frac{0}{1} = 0.0$

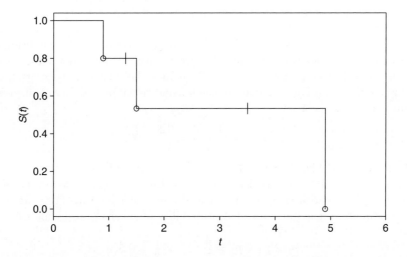

Fig. 8.2. The estimated survival curve from three uncensored (circles) and two censored (crosses) survival times.

dataset includes information from 124 patients, 62 treated with CAPD and 62 with APD. The two Kaplan–Meier estimates for the data are shown in Figure 8.3. The estimated long term (beyond 5000 days, or approximately 14 years) survival rates are similar for the two methods of dialysis, but the CAPD group experienced considerably more early deaths; for example, the estimated five-year survival rates are approximately 0.45 in the CAPD group and 0.75 in the APD group. Figure 8.3 also shows 95% confidence limits on each estimate of $S(t)$, which indicate clearly how the precision of estimation deteriorates as time increases.

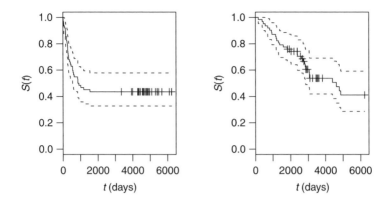

Fig. 8.3. Kaplan–Meier estimates of the survival curve, $S(t)$, for 62 patients receiving CAPD (left-hand panel) and for 62 patients receiving APD (right-hand panel). Dashed lines in both panels indicate 95% confidence limits for the estimated survival curve.

8.3 How long do you expect to live?

At the time of writing (June 2010), life expectancy in the UK is 82.4 years for a woman, but only 78.5 years for a man. However, if you are a man reading this book you must already have survived the rigours of early childhood so, all other things being equal, *your* life expectancy must be longer than 78.5 years. If you are a 60-year-old man, your current life expectancy is in fact 82.3 years. Figure 8.4 explains this pictorially. It shows a hypothetical distribution of lifetimes, with a vertical line added at age 60. For a 60-year-old, we can rule out the possibility that they die before reaching the age of 60, so the relevant distribution is all to the right of the vertical line. To draw a sample from the distribution of lifetimes for people who have already reached age 60, we would draw a sample from the distribution shown in Figure 8.4 but discard any values less than 60. This would yield bigger values (longer lifetimes) on average than would a sample drawn from the original distribution. In more formal language, the distribution confined to the right of the vertical line in Figure 8.4 is the *conditional distribution of lifetime, given that lifetime is at least 60*.

So what should concern you is not the distribution of lifetime, but the conditional distribution of lifetime given *your* current age. Figure 8.5 compares the survival functions, $S(t)$, that correspond to the unconditional and conditional distributions shown in Figure 8.4. Both, necessarily, start at the value $S(0) = 1$, corresponding to certainty, but the conditional survival function maintains this value until age $t = 60$, whereas the unconditional survival function takes values less than 1 for all positive lifetimes.

Now let's take the argument a step further. If you are a 60-year-old man, it may comfort you to know that your remaining life expectancy is not 18.5

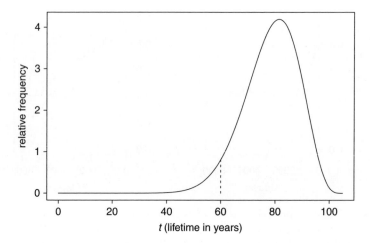

Fig. 8.4. A hypothetical distribution of lifetimes. The average value of samples drawn from this distribution is 78.5 years. The distribution confined to the right of the vertical line is (proportional to) the conditional distribution of lifetime, given survival to age 60. The average value of samples drawn from this distribution is 82.3 years.

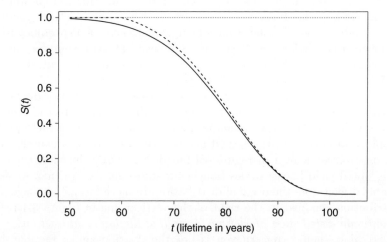

Fig. 8.5. Survival functions for the lifetime distribution shown in Figure 8.4 (solid line) and for the corresponding conditional distribution given survival to age 60 (dashed line). Note that both functions are plotted only for ages greater than 50.

years, but a massive 22.3 years. But you might be more interested to know your current risk of death, i.e., if you are 60 years old, how likely is it that you will die before your 61st birthday? The function that describes your current risk of death at age t is called the *hazard function*, usually written

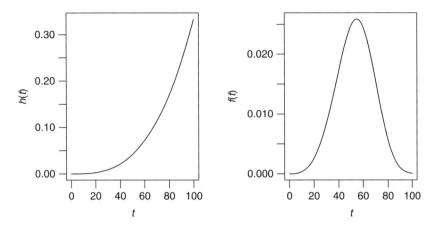

Fig. 8.6. An increasing hazard function $h(t)$ (left-hand panel) and its corresponding lifetime distribution $f(t)$ (right-hand panel).

as $h(t)$. Figures 8.6, 8.7 and 8.8 show three hypothetical hazard functions $h(t)$ and the corresponding distributions, $f(t)$ say, of lifetimes. In the first of these, the hazard (current risk of death) increases with age, which may be a reasonable assumption for human lifetimes in developed countries. In the second, the hazard decreases from birth to age 10 and increases thereafter. This may be a better model for human lifetimes in developing countries where death in early childhood is more common. In the third example, the hazard is constant through life. This implies, paradoxically, that conditional on your surviving to any given age, the distribution of

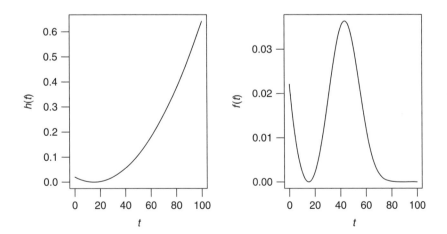

Fig. 8.7. A non-monotone hazard function $h(t)$ (left-hand panel) and its corresponding lifetime distribution $f(t)$ (right-hand panel).

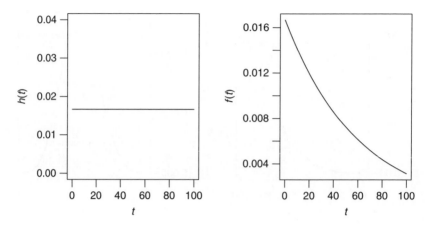

Fig. 8.8. A constant hazard function $h(t)$ (left-hand panel) and its corresponding lifetime distribution $f(t)$ (right-hand panel).

your *remaining* lifetime is the same as it was at birth. This actually does make sense in some contexts. For example, the risk of my being struck by lightning probably does not depend on my age, provided I remain fit enough (and perhaps foolish enough) to insist on taking my daily walk irrespective of the weather. Notice that the relationship between the shape of the hazard function $h(t)$ and its corresponding lifetime distribution $f(t)$ is not at all intuitive.

8.4 Regression analysis for survival data: proportional hazards

In Chapter 6 we described methods for investigating the effects of different experimental treatments on the distribution of a response variable. Our working assumption was that the effect, if any, of a change in treatment would be a simple shift in the distribution of the response. When the response variable is lifetime, this assumption loses much of its appeal. A more natural way to think about the problem is in terms of risk factors, as in newspaper headlines of the form 'new wonder drug reduces risk of heart disease by 50%'. If true, this is equivalent to a change in the hazard function from $h(t)$ to $0.5 \times h(t)$, for any value of t.

In Chapter 7 we extended the methods of Chapter 6 to investigate the effects of multiple design variables and/or explanatory variables on a response, again under the working assumption that their combined effect was to shift the distribution of the response. In the survival analysis setting, a more natural assumption is that their combined effect is to multiply the hazard function by a factor that depends on the values of the design and explanatory variables. A model of this kind is called a *proportional hazards model*. Since its introduction by Cox (1972), the proportional hazards model has become the standard model for survival analysis, especially in medical settings; the paper won for its author the 1990 Kettering prize for Cancer Research.

To understand the proportional hazards model, it helps to think of a specific context, such as the data in Table 8.1. We want to understand how the choice between CAPD and APD, and the patient's age, affect their survival prognosis. We first choose a set of standard circumstances, say a 50-year-old patient being treated with CAPD. Let $h_0(t)$, called the *baseline hazard function*, be the hazard function for such a patient. Now, the proportional hazards model assumes that the hazard for a patient being treated by CAPD, is

$$h(t) = h_0(t) \exp\{\beta \times (\text{age} - 50)\},$$

and for a patient being treated by APD, is

$$h(t) = h_0(t) \exp\{\alpha + \beta \times (\text{age} - 50)\}.$$

We can combine these into a single expression by defining a design variable 'method' that takes the value 0 for CAPD, 1 for APD. Then,

$$h(t) = h_0(t) \exp\{\alpha \times \text{method} + \beta \times (\text{age} - 50)\}. \tag{8.1}$$

Note that we could also consider the effects of treatment method and age in two separate models,

$$h(t) = h_0(t) \exp\{\alpha \times \text{method}\} \tag{8.2}$$

and

$$h(t) = h_0(t) \exp\{\beta \times (\text{age} - 50)\}, \tag{8.3}$$

respectively. However, for reasons that we will explain in the next section, there is no reason to suppose that the two approaches will give the same, or even similar answers.

One reason for the popularity of the proportional hazards model is that it allows you to investigate the effects of design variables and explanatory variables without having to specify the form of the baseline hazard function. In the current example, this means that you can reach a conclusion on the relative merits of CAPD and APD without having to model $h_0(t)$; put another way, any conclusions you reach about the choice of treatment apply irrespective of the actual form of $h_0(t)$.

8.5 Analysis of the kidney failure data

We now use the proportional hazards model to analyse the kidney failure data, and explain how to interpret the results of the analysis. As in

Chapter 7, we will show the output as it is produced by the R software. However, most software implementations for proportional hazards modelling will give very similar output, and it should be reasonably straightforward to adapt the explanation given here to whatever software you are using. And, as always, you can reproduce the analysis using the data and R code provided on the book's website if you wish.

First, we consider whether the suggestion from Figure 8.3 that patients have a better survival prognosis when treated with APD rather than with CAPD is backed up by formal inference. To do this, we fit a proportional hazards model with treatment type as a two-level factor. The relevant part of the R output is shown below.

```
  n= 124
          coef exp(coef) se(coef)      z     p
method   -0.48     0.619     0.25  -1.92 0.055

        exp(coef) exp(-coef) lower .95 upper .95
method      0.619       1.62     0.379      1.01
```

The first line of output confirms that the data cover 124 patients. The next two lines tell us that the fitted effect (coef, an abbreviation of 'coefficient') of the two-level factor method is -0.48. This is the estimated value of the parameter α in equation (8.2). The fact that the estimate is negative implies that the estimated hazard is lower under APD than under CAPD. The next column, labelled exp(coef), is simply the exponential of coef, with value 0.619. This gives exactly the same information as coef, but in a more easily interpretable form: the hazard under APD is estimated to be 0.62 times the hazard under CAPD, i.e., APD conveys a 38% reduction in risk. The last three columns give us the information we need to test the hypothesis that there is no difference between the two treatment arms. The test statistic (z) is the ratio of $\hat{\alpha}$ to its standard error (se(coef)), and the final column (p) gives the p-value of the test, which does not quite reach the conventional 5% (0.05) level of significance. Rigid adherence to convention would force us to declare the analysis inconclusive on the key question of which, if either, version of peritoneal dialysis has the better survival prognosis. The last two lines of output repeat the estimated hazard ratio (APD relative to CAPD, exp(coef)) and, in case you prefer to express the hazard ratio as CAPD to APD rather than the other way round, its reciprocal (exp(-coef)). More usefully, the last two columns give the 95% confidence interval for the hazard ratio (but if you prefer to express these as CAPD to APD, you have to calculate the reciprocals yourself!).

Clinicians know perfectly well that younger patients are more likely to survive longer than older patients. Is our comparison of the two treatment arms affected if we allow for an age effect? To answer this question we fit the model (8.1) and obtain the following output.

```
n= 124
         coef  exp(coef) se(coef)     z      p
method -0.9908    0.371  0.26418  -3.75  1.8e-04
age     0.0566    1.058  0.00838   6.76  1.4e-11

       exp(coef) exp(-coef) lower .95 upper .95
method    0.371      2.693     0.221     0.623
age       1.058      0.945     1.041     1.076
```

The format of the output is essentially the same as before, except that now we are given information about the estimated effects of both treatment (`method`) and age. Now, the confidence interval for the hazard ratio (`exp(coef)`) between APD and CAPD comfortably excludes the neutral value one. This gives a strong indication that patients of the same age have a better survival prognosis under APD than under CAPD. Also, the confidence interval for the age effect indicates that, with 95% confidence, the effect of each extra year of age is to increase the hazard by a factor of between 1.041 and 1.076 (4.1 to 7.6%).

Why did the two sets of results differ with respect to the comparison between APD and CAPD? One explanation for the discrepancy is that there are fewer young patients in the APD arm than in the CAPD arm. This would have been very unlikely had the data arisen from a randomized trial. But they were collected in routine clinical practice, and it may have been that the younger patients were more likely to be treated by CAPD, for example because it was thought to be better suited to their lifestyle. In this case, a comparison that fails to take account of the age effect is, arguably, unfair to APD.

A variable in a dataset whose effect is not of direct interest, but which is statistically related to one that *is* of direct interest, is called a *confounder*. Whenever possible, an analysis of such data should adjust for the effect of the confounder before reaching any conclusions regarding the variable of direct interest. This raises the uncomfortable possibility that there may be *unmeasured* confounders for which, by definition, we can make no such adjustment. This is one reason why the randomized trial is considered to be a gold standard, and why some statisticians are sceptical of *any* findings drawn from observational studies. Our view is that the latter position is far too harsh. Certainly, findings from observational studies should be treated cautiously, but if their preliminary findings are backed up by subsequent studies, and more importantly by a scientific explanation, they become robust. Nobody ever has, or ever will, conduct a randomized trial of the relationship between smoking and the risk of lung cancer.

8.6 Discussion and further reading

The proportional hazards model is probably the most widely used approach to analysing survival data, but it is not the only option. Another popular model is the *accelerated life* model. Here, the key assumption is that the combined effect of design variables and explanatory variables is to change the speed of the biological clock. So, for example, in the context of our renal data, if the survival function for a 50-year-old patient being treated with CAPD is $S(t)$, the accelerated life model would assume that the survival function for a 50-year-old on APD takes the form $S(\beta t)$, where the parameter β measures the acceleration (or deceleration if $\beta < 1$) effect of APD by comparison with CAPD.

A third option is to deal with the fact that lifetimes are necessarily positive by transforming the data, for example to logarithms of survival times, and using the methods described in Chapters 6 and 7. This provides a particularly simple approach in the absence of censoring.

For readers who would like to know more about survival analysis, good introductory accounts include Collett (2003) or Cox and Oakes (1984). More advanced texts include Kalbfleisch and Prentice (2002) and Lawless (2003).

For a brief account of how the steady accumulation of evidence from observational studies established the link between smoking and lung cancer, see Chapter 19 of Anderson (1989).

9
Time series analysis: predicting fluctuations in daily maximum temperatures

9.1 Weather forecasting

Figure 9.1 shows the daily maximum temperatures recorded at the Hazelrigg research station, near Lancaster, over a one-year period from 1 September 1995 to 31 August 1996. The plot shows some predictable features: strong seasonal variation, together with relatively small day-to-day fluctuations about the prevailing seasonal pattern. The seasonal pattern is a reflection of the *climate* in this part of the world, whilst the fluctuations reflect changes in the *weather*. Another way to express this distinction is that the seasonal pattern is an example of *deterministic* or *systematic* variation, whilst the day-to-day fluctuations are *stochastic*, or *random* (climate is what you expect, weather is what you get). How can we use data of the kind shown in Figure 9.1 to make temperature forecasts?

9.2 Why do time series data need special treatment?

We denote an observed time series by $y_t : t = 1, \ldots, n$, where n is the length of the series and y_t denotes the value of the series recorded in time unit t. In our motivating example, $n = 365$, the time unit is one day, $t = 1$ corresponds to 1 September 1995 and each y_t is a temperature in degrees Celsius.

A common feature of all of the statistical methods that we have discussed so far is that they assume *independent replication* between experimental units. In a designed experiment, and as discussed in Chapter 5, careful attention to experimental technique can make the independence assumption reasonable. When dealing with time series data, independence between observations on successive time units cannot be assumed, and in practice often does not hold. In our motivating example, we might expect periods of several successive warmer-than-average days to be followed by a run of cooler-than-average days.

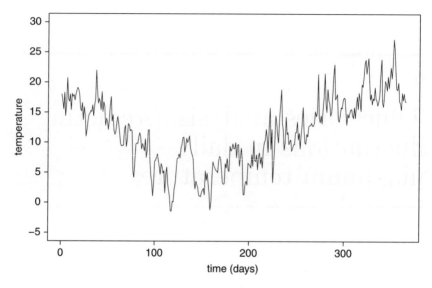

Fig. 9.1. Daily maximum temperatures, in degrees Celsius, recorded at Hazelrigg field station from 1 September 1995 to 31 August 1996.

A consequence of the lack of independence between the values of y_t is that methods based on the assumption of independent replication are invalid. For example, the usual formula (6.1) for the standard error of a mean can be wrong by a factor that could be as large as the square root of the sample size, n. To avoid reaching incorrect conclusions from time series data, special methods are needed that can make allowance for weak or strong dependence as appropriate.

9.3 Trend and seasonal variation

To understand the behaviour of the processes that generate an observed time series, we need to decompose the variation in the data into two or more components. The simplest such decomposition is into *deterministic* and *stochastic* variation. Both trend and seasonality refer to aspects of the deterministic component.

The *trend* in a time series is its average value as a function of time. Strictly, this assumes that, at least in principle, the process that generated the observed time series can be replicated. For experimentally generated time series, such replication is feasible in practice, and in these circumstances the easiest way to think about the trend is that it represents what a series of averages of observed values at each time point would look like if the experiment were repeated indefinitely under identical conditions. For observational series like the Hazelrigg temperature data, this is not physically possible: there will never be another 1 September 1995. Nevertheless, if we are prepared to imagine a set of parallel universes in which the same laws of nature prevail, we might be prepared to predict that in every one of them summer will tend to be warmer than winter, but not to predict the precise temperature on a given date.

The authors' view of *seasonality* is that it forms part of the trend, specifically that part, if any, that is reproduced cyclically over time. For our temperature data, the seasonality in the data coincides with the everyday use of the word, but in other contexts the length, or *period* of a cycle may take some other value, for example a 24-hour cycle in hourly temperature readings.

Once we have removed the deterministic component of variation from a time series, we assume that the remaining variation can be explained by appealing to the laws of probability. The term *stochastic* means 'governed by the laws of probability'. The simplest possible model of the stochastic component of variation in a time series is the so-called *white noise* model, consisting of a sequence of independent random fluctuations about the underlying trend. Very often this model fails adequately to describe the data, but it provides a starting point for building more flexible and realistic models.

The astute reader will have spotted that our brief appeal to the idea of parallel universes to justify stochastic modelling of observational time series data lacks rigour. Strictly, if we cannot replicate the natural process we wish to study, so that we can only observe a single time series, there is no absolute criterion through which we can disentangle the deterministic and stochastic components of variation in the series. Instead, we have to make more or less subjective judgements, ideally informed by scientific knowledge of the natural processes that generate the data, about what aspects of the data we should model *as if* they were stochastic.

By way of illustration, Figure 9.2 shows a time series of weekly black smoke pollution levels (concentration in air, in parts per million) in the city of Newcastle upon Tyne, over the ten-year period 1970 to 1979 inclusive. The data show a clear, and unsurprising, seasonal pattern in the everyday sense, with higher pollution levels in winter than in summer, and a long-term decreasing trend as a direct result of the steady decline in domestic coal-burning during the 1970s. However, the magnitude of the seasonal variation, and the time of year at which pollution is highest, vary from year to year. Again, the scientific context makes this unsurprising: black smoke results in part from the burning of coal for domestic heating, which at the time these data were recorded was widespread in the city. Pollution levels therefore reflect changes in temperature and, as we all know from experience, not all winters are equally cold and the coldest time of the year does not always fall on the same date. The model fitted to the data, which is shown in Figure 9.2 as a smooth, dashed curve, captures this behaviour by including a cosine wave with an annual period, $A\cos(2\pi t/52 + P)$, where

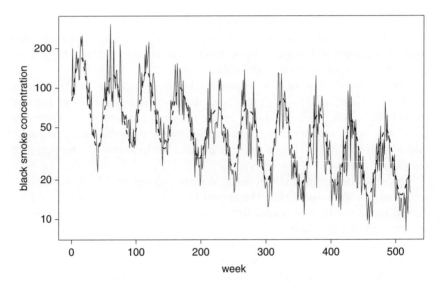

Fig. 9.2. Weekly black smoke levels (ppm) in the city of Newcastle upon Tyne (black line) and a fitted stochastic model for the underlying trend (dashed line).

t is time in weeks, A is the *amplitude* and P the *phase* of the seasonal variation. Crucially, the model also allows the values of A and P to fluctuate stochastically over time. A detailed description of these data and the fitted model can be found in Fanshawe *et al.* (2008).

9.4 Autocorrelation: what is it and why does it matter?

In Chapter 4, we introduced the idea of correlation as a numerical measure of the strength of linear association between pairs of measurements. One way to construct pairs of measurements from a time series is to consider a time separation, or *lag*, of k time units and define pairs (y_t, y_{t-k}) : $t = k+1, \ldots, n$. The correlation between such pairs is called the lag-k *autocorrelation* of the series.

Figure 9.3 shows a set of scatterplots of the lagged pairs (y_t, y_{t-k}) of Hazelrigg temperatures, for $k = 1, 2, 3, 4$. All four panels show a very strong, positive association. This is neither surprising nor particularly interesting. It only confirms that temperature varies systematically over the year, hence pairs of temperatures separated by only a few days will be relatively similar, i.e., positively associated, simply by virtue of their closeness in time. Hence, our first observation about autocorrelation analysis is that it is of little or no value unless any deterministic trend in the data has been identified and removed.

A potentially interesting question is not whether the temperature series itself shows evidence of autocorrelation, but whether the time series, z_t

AUTOCORRELATION: WHAT IS IT AND WHY DOES IT MATTER?

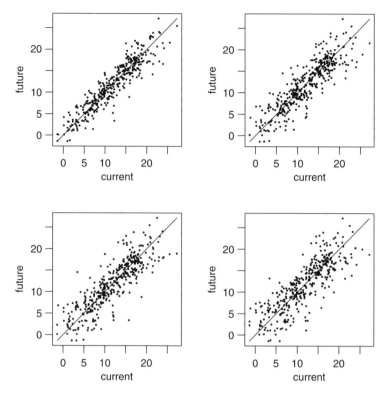

Fig. 9.3. Scatterplots of lagged pairs of daily maximum temperatures, in degrees Celsius, recorded at Hazelrigg field station from 1 September 1995 to 31 August 1996. Time lags are 1, 2, 3 and 4 days (top-left, top-right, bottom-left and bottom-right panels, respectively).

say, of fluctuations about the seasonal trend are autocorrelated. Figure 9.4 gives a positive answer. To produce this diagram, we calculated each z_t as the corresponding temperature, y_t, minus the value on day t of a simple sine–cosine wave fitted to the data by defining $x_{1t} = \cos(2\pi t/366)$, $x_{2t} = \sin(2\pi t/366)$ and fitting the linear model

$$y_t = \alpha + \beta_1 x_{1t} + \beta_2 x_{2t} + Z_t$$

as described in Sections 7.5 and 7.6. See also Section 9.5 below. Notice also that the autocorrelation in the series z_t is weaker at longer lags.

A useful device to summarize the pattern of autocorrelation in a time series is a plot of the lag-k autocorrelation, r_k, against k. Such a plot is called a *correlogram*. Conventionally, the r_k are calculated after subtracting the sample mean, $\bar{y} = (\sum_{t=1}^{n} y_t)/n$, from each y_t. Hence,

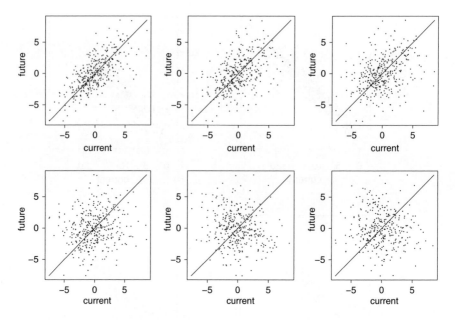

Fig. 9.4. Scatterplots of lagged pairs of daily maximum temperature residuals. Time lags are 1, 2, 3 days (top row, left to right) and 7, 14 and 28 days (bottom row, left to right).

$$r_k = g_k/g_0, \qquad (9.1)$$

where, for each value of $k \geq 0$,

$$g_k = \sum_{t=k+1}^{n} (y_t - \bar{y})(y_{t-k} - \bar{y}). \qquad (9.2)$$

As already noted, the correlogram should only be calculated after any trend in the time series has been removed. Figure 9.5 shows correlograms of the Hazelrigg temperature series before and after trend removal using the simple sine–cosine wave model for the trend. The correlogram of the raw data series is distorted by the underlying trend, which induces a positive correlation at all of the plotted lags. This is not wrong on its own terms, but it tells us nothing new: a similar effect would be observed in the correlogram of any time series with a smooth time trend. The correlogram of the residual series is more interesting, because it indicates that the series of temperature fluctuations about the trend is positively autocorrelated, and that the magnitude of this autocorrelation appears to decay smoothly towards zero with increasing lag. This reflects a well-known feature of the UK climate, namely that relatively warm or cool spells tend to persist over

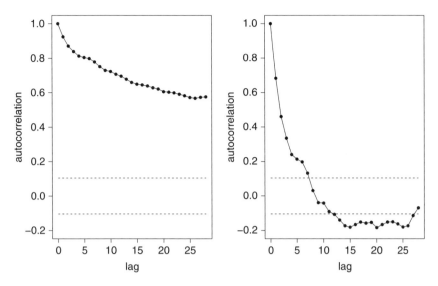

Fig. 9.5. Correlograms of maximum daily temperatures (left-hand panel) and of residuals (right-hand panel). Dashed horizontal lines are set at values $\pm 2/\sqrt{365}$.

several days, but rarely over several weeks. In Section 9.5 we will show how we can exploit this in order to make weather forecasts.

The dashed horizontal lines on the correlograms are set at values $\pm 2/\sqrt{n}$. These represent approximate 95% point-wise tolerance limits under the assumption that the process generating the data is a sequence of independent fluctuations about a constant mean. An immediate conclusion is that the autocorrelations at small lags k are statistically significant (from zero, at the conventional 5% level). More interestingly, the smooth decay pattern in the autocorrelation with increasing lag suggests that it should be possible to build a descriptive model for the complete set of autocorrelations, r_k, rather than to regard them as taking arbitrarily different values.

9.5 Prediction

A common goal in time series analysis is to predict the future. How can we use the Hazelrigg temperature data to forecast tomorrow's, or next week's, or next month's temperature?

Let's first consider long-term weather forecasting. If we want to predict the temperature one month ahead, knowing whether today is relatively warm or cold for the time of year is not very helpful; probably the best we can do is predict that the temperature next month will be at its seasonal average (of course, professional meteorologists can do better than this by examining much more extensive national and global data than we have at our disposal).

In contrast, if today is warmer than average for the time of year, most of us would predict that tomorrow will also be warmer than average, and conversely. The intermediate problem of predicting next week's temperature is trickier. Is one week ahead a short-term or a long-term prediction? Clearly, it lies somewhere between the two, but where exactly? This is where autocorrelation can help us. For a prediction of the temperature k days ahead, the stronger the lag k autocorrelation, the more we would expect today's stochastic variation about the seasonal average to persist k days ahead of today. Put more formally, if we are given today's temperature, the uncertainty in predicting temperature k days ahead, expressed as the variance of the prediction, is reduced by a factor of $1 - r_k^2$.

To put this into practice, we need to build a model for the Hazelrigg data. For the deterministic component, we use a cosine wave, $\alpha \cos(2\pi t/366 + \phi)$. This differs from the model for the black smoke data in two ways. Firstly, because the data are recorded daily rather than weekly, 366 replaces 52 so as to preserve the required annual cycle (note that 1996 was a leap year). Secondly, because we only have one year's data, the amplitude and phase are treated as fixed parameters rather than as stochastically varying over time. Fitting the model turns out to be easier if we assume, temporarily, that the fluctuations about the seasonal trend are a white noise series Z_t, i.e., values of Z_t on different days t are independent (this assumption is palpably false, but bear with us). We write Y_t for the temperature on day t and μ for the year-long average temperature. Also, by using the trigonometric identity that $\cos(A + B) = \cos(A)\cos(B) - \sin(A)\sin(B)$, we can re-express our provisional model for the data as

$$Y_t = \mu + \beta_1 \cos(2\pi t/366) + \beta_2 \sin(2\pi t/366). \tag{9.3}$$

The new parameters, β_1 and β_2, in (9.3) relate to the amplitude, α, and phase, ϕ, as follows:

$$\alpha = \sqrt{(\beta_1^2 + \beta_2^2)}, \tag{9.4}$$

$$\phi = \tan^{-1}(\beta_2/\beta_1). \tag{9.5}$$

The advantage of the re-expression (9.3) is that the values $x_1 = \cos(2\pi t/366)$ and $x_2 \sin(2\pi t/366)$ do not depend on any unknown quantities, hence (9.3) is a disguised *linear* regression model,

$$Y_t = \mu + \beta_1 x_1 + \beta_2 x_2 + Z_t \tag{9.6}$$

of the kind discussed in Chapter 7 and the parameters μ, β_1 and β_2 can be estimated using standard regression software.

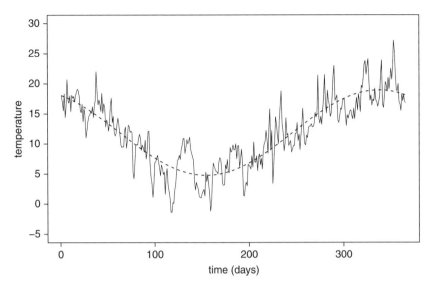

Fig. 9.6. Daily maximum temperatures, in degrees Celsius, recorded at Hazelrigg field station from 1 September 1995 to 31 August 1996 (solid line) and fitted harmonic regression model (dashed line).

We used the methods described in Sections 7.5 and 7.6 to fit the model, obtaining estimates $\hat{\mu} = 11.83$, $\hat{\beta}_1 = 6.27$ and $\hat{\beta}_2 = -3.27$. The corresponding values of the physically more natural quantities of amplitude and phase are $\hat{\alpha} = 7.07$ and $\hat{\phi} = -0.48$ radians. Figure 9.6 shows the data with the fitted model. The fitted curve captures the seasonal pattern quite well. A more subtle question is whether the daily fluctuations about the curve agree with the white noise assumption for Z_t in (9.6). As we know already, the correlogram of the residuals, $\hat{Z}_t = Y_t - \hat{\mu} - \hat{\beta}_1 \cos(2\pi t/366) - \hat{\beta}_2 \sin(2\pi t/366)$, clearly suggests not.

A simple way to model the autocorrelation in these data is to think of tomorrow's fluctuation about the seasonal trend as a compromise between today's fluctuation and an independent stochastic perturbation. Expressed formally, we now assume that the fluctuations Z_t follow the model

$$Z_t = \rho Z_{t-1} + \epsilon_t, \qquad (9.7)$$

where ρ represents the correlation between Z_t and Z_{t-1}, and must therefore take a value between -1 and $+1$, and ϵ is a white noise sequence. As with the model for the seasonal trend, we can re-express (9.7) as a linear regression model by defining, for each day t, the response to be today's temperature residual and the explanatory variable to be yesterday's temperature residual. To fit the model, we then constrain the intercept of the linear fit to be zero. This gives the estimate $\hat{\rho} = 0.684$. Figure 9.7 compares

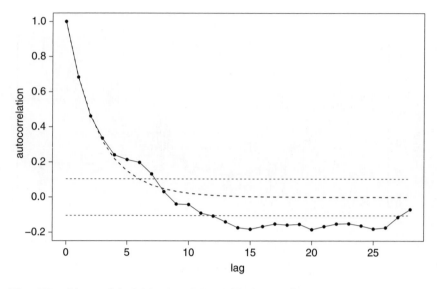

Fig. 9.7. Observed (solid line) and fitted (dashed line) autocorrelation functions for time series of maximum daily temperatures.

the fitted and observed correlograms. The model captures the general shape of the autocorrelation function well. The negative observed autocorrelations at large lags are incompatible with the model, even allowing for sampling variation as expressed by the horizontal dashed lines set at plus and minus $2/\sqrt{366}$ on Figure 9.7. A partial explanation is that the plus and minus $2/\sqrt{366}$ limits provide only approximate guidelines that are less reliable at large than at small lags, and strictly only applicable to assessing departures from a completely uncorrelated series. The sine–cosine model for the seasonal trend is also likely to be no more than an approximation. A cautious conclusion would be that the model gives a reasonably good, if imperfect, fit to the observed pattern of autocorrelation in the data.

Now, suppose that on day t we want to predict the temperature k days ahead. The prediction proceeds in stages as follows. Firstly, we compute the estimated average temperature on each day of the year,

$$\hat{\mu}_t + \hat{\beta}_1 \cos(2\pi(t)/365) + \hat{\beta}_2 \sin(2\pi(t)/365).$$

Secondly, we compute the corresponding residuals,

$$Z_t = Y_t - \hat{\mu}_t.$$

Strictly, this only gives an approximation to the correct values of Z_t because we have ignored the uncertainty in our estimation of μ_t, but it is close enough for our purposes. Next, we repeatedly use (9.7) as follows. Since equation (9.7) holds for any time t, we can write

$$Z_{t+2} = \rho Z_{t+1} + \epsilon_{t+2}$$
$$= \rho(\rho Z_t + \epsilon_{t+1}) + \epsilon_{t+2}$$
$$= \rho^2 Z_t + \rho \epsilon_{t+1} + \epsilon_{t+2}. \tag{9.8}$$

Now, because both ϵ_{t+2} and ϵ_{t+1} have mean zero and are independent of Z_t, the data are of no help to us in predicting whether their actual values will be positive or negative, and a sensible procedure is to predict them as zeros. Then, (9.8) implies that a sensible predictor for Z_{t+2} at time t is $\rho^2 Z_t$. Applying this argument a second time shows that a sensible predictor of Z_{t+3} at time t is $\rho^3 Z_t$, and so on. This gives us the general rule that, to predict temperature k days ahead, we add the estimated average temperature and the predicted fluctuation about the average, hence

$$\hat{Y}_{t+k} = \hat{\mu}_{t+k} + \rho^k (Y_t - \hat{\mu}_t). \tag{9.9}$$

To apply this rule to the Hazelrigg data, we need to estimate the parameter ρ. Recall that ρ represents the correlation between successive fluctuations Z_t. Hence, our estimate for ρ is the lag-1 autocorrelation of the residual series, as shown in Figure 9.5, namely $\hat{\rho} = 0.683$.

How does this prediction rule work in practice? In particular, how much, if at all, better is it than other, simpler rules? Ideally, prediction algorithms should be evaluated on data other than those used to fit the associated statistical models. Here, in the absence of any additional data, we compare three prediction rules by applying them retrospectively to the Hazelrigg data. The rules are:

(1) $\hat{Y}_{t+k} = \hat{\mu}_{t+k} + \hat{\rho}^k (Y_t - \hat{\mu}_t)$ (model-based)
(2) $\hat{Y}_{t+k} = \hat{\mu}_{t+k}$ (deterministic)
(3) $\hat{Y}_{t+k} = Y_t$ (random walk)

A statistician would expect rule 1 to perform best, because it is derived from a model that has been shown to fit the data well. Rule 2 is labelled 'deterministic' because it uses only our best estimate of the systematic component of the data and ignores the effect of stochastic fluctuations. Rule 3 says that the future will be very like the present, which for small values of k may not be a bad approximation to the truth. From a statistical modelling perspective, it would be optimal if the underlying process followed a random walk, defined as $Y_{t+1} = Y_t + Z_t$, where Z_t is white noise. In fact, this very simple model works surprisingly well in many settings.

To compare the three rules, we use the following *mean square error* criterion. Suppose that at time t we wish to predict the value, y_{t+k}, of the temperature k days later. A prediction rule gives the predicted value $\hat{y}_{t+k}$. Then, the *mean square error* of the rule is

Table 9.1. Mean square error, MSE, evaluated for each of three prediction rules and for forecast lead times $k = 1, 2, 3, 7, 14, 28$. See text for definition of rules 1, 2 and 3.

Rule	Forecast lead time (days)					
	1	2	3	7	14	28
1	4.03	5.96	6.73	7.49	7.75	7.86
2	7.56	7.57	7.58	7.60	7.73	7.86
3	4.78	8.15	10.07	13.20	17.86	16.44

$$MSE = \left\{ \sum_{t=1}^{n-k} (y_{t+k} - \hat{y}_{t+k})^2 \right\} / (n-k).$$

Table 9.1 tabulates the values of MSE for each of the three rules, and for each of the *forecast lead times* $k = 1, 2, 3, 7, 14, 28$. As expected, rule 1 gives the best performance overall. Rule 2 performs relatively poorly at small lead times k, but at large lead times performs essentially as well as rule 1. The explanation for this is that rule 1 differs from rule 2 by taking account of the lag-k autocorrelation in the data, whereas rule 2 assumes implicitly that this autocorrelation is zero, which is not a bad approximation for large values of k. The fitted autocorrelation used by rule 1 is $r_k = 0.683^k$. Hence, $r_1 = 0.683$, $r_2 = 0.466$, $r_3 = 0.318$ and so on until $r_{28} = 0.000023 \approx 0$. Rule 3 outperforms rule 2 at lag $k = 1$ and has comparable performance at lag $k = 2$, but its performance deteriorates thereafter, because it uses a model which assumes, incorrectly, that the strong autocorrelation at small lags persists to large lags.

9.6 Discussion and further reading

Time series analysis is one of the more specialized branches of statistics. Introductory books include Chatfield (1996) and Diggle (1990).

More advanced statistical methods for time series data than those covered here have much in common with signal processing methods used in physics and electrical engineering. Two examples are *spectral analysis* and *Kalman filtering*.

The central idea in spectral analysis is to represent a time series as a superposition of sine–cosine waves at different frequencies. This is a physically natural approach for many electrical series, but also has an empirical justification in that *any* series of length n can be approximated arbitrarily closely by at most $n/2$ sine–cosine waves and often in practice

DISCUSSION AND FURTHER READING

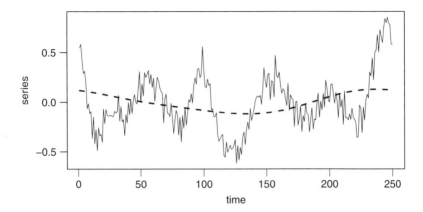

Fig. 9.8. A simulated time series (solid line) and an approximation to it (dashed line) obtained by the superposition of two sine–cosine waves (dashed line).

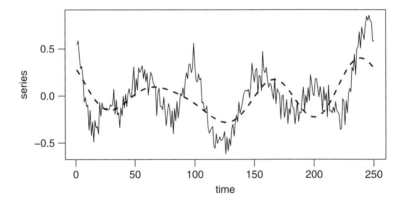

Fig. 9.9. The same simulated time series as in Figure 9.8 (solid line) and an approximation to it (dashed line) obtained by the superposition of four sine–cosine waves.

by many fewer. As an example, Figures 9.8, 9.9 and 9.10 show a time series of length $n = 250$ and successive approximations to it, each of which is the sum of r sine–cosine waves, $r = 2, 4, 8$. The first approximation is very poor. The second does a slightly better job, whilst the third succeeds in capturing the main features of the variation in the observed time series.

The Kalman filter derives its name from the pioneering work of R. E. Kalman (Kalman, 1960; Kalman and Bucy, 1961). The key idea is to represent a time series as the sum of a *signal*, which we wish to detect, and *noise*, which is hiding the signal from us. The Kalman filter is an algorithm for finding the best estimate of the underlying signal in real time, i.e., the

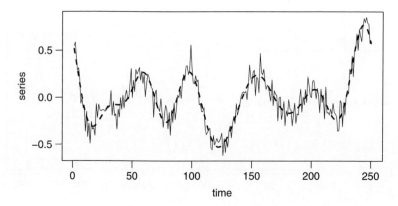

Fig. 9.10. The same simulated time series as in Figure 9.8 (solid line) and an approximation to it (dashed line) obtained by the superposition of eight sine–cosine waves.

estimate of the signal is updated efficiently whenever a new observation becomes available. For an introductory account, see for example Meinhold and Singpurwalla (1983). In Figure 9.2, the smooth curve showing the trend and seasonal variation in black smoke concentrations was estimated using the Kalman filter.

10
Spatial statistics: monitoring air pollution

10.1 Air pollution

Figure 10.1 is a day-time view of central London in December 1952 during a smog (an amalgam of 'smoke' and 'fog'). Conditions like this were a feature of the UK's winter urban environment at the time. They were caused by an unhappy combination of weather conditions and pollution from coal-burning and traffic fumes, and were somewhat affectionately known as 'pea soupers'. But the exceptionally severe smog of winter 1952–1953 led to a dramatic increase in the death rate, demonstrating what is now well known – that polluted air can have serious adverse effects on human health. Consequently, throughout the developed world air quality is now monitored to enable corrective measures to be implemented if the concentrations of any of a range of pollutants in the air exceed agreed, and regulated, safe levels.

The most obvious way to measure air quality is by physical and chemical analysis of directly sampled air. However, this is an expensive process and as a result monitoring networks tend to be spatially sparse. An example is shown in Figure 10.2, taken from Fanshawe *et al.* (2008). That paper reports a study of black smoke pollution (solid air-suspended particles with diameter of at most $4\,\mu m$) in and near the city of Newcastle upon Tyne at various times over the period 1962 to 1992. Although Figure 10.2 shows 25 monitor locations, not all were in operation at any one time, and by the end of the study period only three remained active.

This has led scientists to consider other, less direct but perhaps more cost-effective, technologies for monitoring pollution, including biomonitoring. The idea of this is that instead of measuring the concentrations of particular pollutants directly in sampled air, we can exploit the ability of plants to take up pollutants, whose concentrations can then be determined by chemical analysis of plant tissue. Figure 10.3 shows data from a biomonitoring study of this kind. The positions of the circles on the map

Fig. 10.1. Central London during a severe smog.

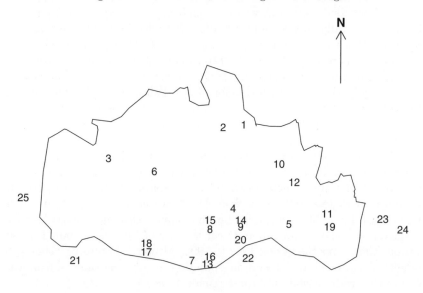

Fig. 10.2. Locations of black smoke monitoring stations in the city of Newcastle upon Tyne. Individual monitors were active over various time periods between 1961 and 1992. Monitors 21 to 25 were located just outside the current city boundary.

denote a set of 63 locations in Galicia, northwestern Spain, at which samples of a particular species of moss were taken, whilst their radii are proportional to the lead concentration determined by chemical analysis of the sample.

Galicia is bounded by the neighbouring province of Asturias to the east, by the river Minho on the border with Portugal to the south, and by the Atlantic Ocean to the west and north. More detailed information on the data collection and scientific context are given in Fernández, Rey and Carballeira (2000). Figure 10.3 gives a strong suggestion of spatial variation in lead concentrations that could not easily be described by simple north–south or east–west trends. The biomonitoring technology has allowed data to be collected from many more locations than would have been feasible using direct air sampling devices. A primary goal in analysing these data is to construct a Galicia-wide map of pollutant concentrations to inform environmental planning and regulation. This requires us to use the data at the sampled locations to predict pollutant concentrations at unsampled locations.

In Chapter 9 we used the term *predict* to mean making an educated guess of the future value of some phenomenon of interest. Here, we use the term in a more general sense, to mean making an educated guess

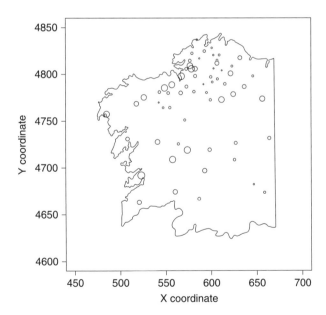

Fig. 10.3. Galicia lead pollution data. Each circle has its centre at the sampling location and its radius proportional to the measured lead concentration. An approximation to the boundary of Galicia is shown to set the data in context, but otherwise plays no part in the analysis of the data.

of the value of any unobserved quantity of interest, without necessarily any connotations of time. In other words, a forecast is a prediction but a prediction may or may not be a forecast.

10.2 Spatial variation

One way to describe the spatial variation in a set of data is to look for spatially referenced explanatory variables and apply the ideas in Chapter 7 to build a suitable regression model. When explanatory variables are not available, or do not adequately explain the observed spatial variation, we adopt a statistical analogue of what is sometimes called the 'first law of geography', which is that close things tend to be similar. Expressed in statistical language, this 'law' states that measured values from near-neighbouring locations will tend to be positively correlated, and that the correlation weakens with increasing separation distance. This principle is at the heart of a methodology for spatial interpolation and smoothing called *geostatistics*; the name is an acknowledgement that the methodology was originally developed in the mining industry to enable prediction of a proposed mine's potential yield, based on the results from exploratory ore samples.

10.3 Exploring spatial variation: the spatial correlogram

Suppose that we have measured values Y_i at spatial locations $x_i : i = 1, \ldots, n$. We wish to explore how the correlation between a pair of measured values depends on the distance between their corresponding locations, and perhaps on their orientation.

If the locations form a regular lattice, we can do this by a straightforward extension of the correlogram for a time series, as discussed in Chapter 9. Suppose, for illustration, that the locations form a unit square lattice. Then, for any *spatial lag*, (r, s) say, we can calculate the correlation between the measured values on the original lattice and on a superimposed copy that has been shifted by r and s units in the horizontal and vertical directions, respectively. This is shown in Figure 10.4, where $r = 3$ and $s = 5$. The collection of correlations obtained in this way over a range of values of r and s is called the *spatial correlogram*.

The data from the wheat uniformity trial that we discussed in Chapter 5 are an example of lattice data. Figure 10.5 shows the spatial correlogram of these data, calculated at spatial separations up to 10 plots in each coordinate direction. The plot gives an impression of a series of parallel ridges of relatively high positive correlations (lighter shades of grey). This suggest a quasi-cyclic pattern of variation, which is thought to be a residual effect of ancient ridge-and-furrow ploughing of the field; for further discussion, see McBratney and Webster (1981). Note also that the correlation at spatial

EXPLORING SPATIAL VARIATION: THE SPATIAL CORRELOGRAM   145

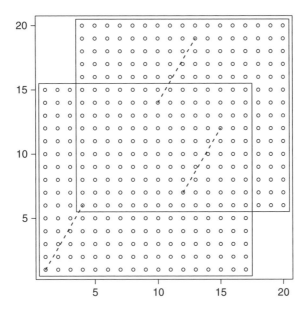

Fig. 10.4. Calculating the spatial correlogram at spatial lags $r = 3$ and $s = 5$. The correlation is calculated from corresponding pairs of measured values on the original and shifted lattices; three such pairs are indicated by the dashed lines.

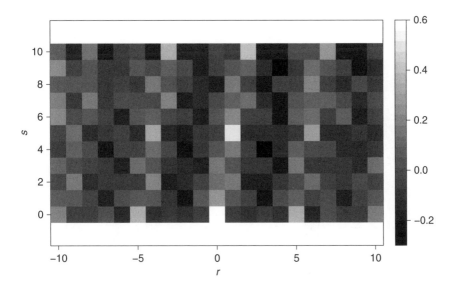

Fig. 10.5. The spatial correlogram of the wheat uniformity trial data at spatial lags up to 10 in each coordinate direction

lag $(0,0)$ is necessarily equal to 1, and is plotted as a white square on Figure 10.5.

If a spatial correlogram does not suggest any directional effects such as are apparent in the wheat uniformity trial data, a useful simplification is to compute correlations between pairs of values at locations separated by a given distance, irrespective of their orientation. To do this for the wheat uniformity data would be misleading, as it would hide a potentially important directional aspect of the spatial variation in wheat yields.

10.4 Exploring spatial correlation: the variogram

An intuitive way to think about spatial correlation is the following. In a set of data with positive spatial correlation at spatial separations r and s in the two coordinate directions, pairs of measurements separated by r and s will, on average, be closer to each other in value than would a pair of randomly sampled measurements. A way of expressing this numerically is to calculate squared differences between all possible pairs of measured values in the data, and average these squared differences at each pair of spatial separations. A plot of average squared differences against spatial separations is called the *variogram* of the data. Conventionally, the variogram is defined as one half of the average squared difference at each spatial separation (which, surprisingly enough, leads some authors to call it the semi-variogram!).

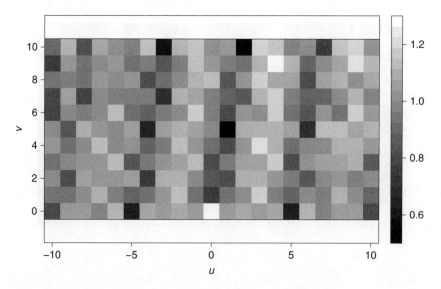

Fig. 10.6. The variogram of the wheat uniformity trial data at spatial lags up to 10 in each coordinate direction.

Figure 10.6 shows the variogram of the wheat uniformity trial data, scaled by dividing each value by the sample variance of the data. Notice how this plot is essentially an inverted version of the spatial correlogram, i.e., the variogram takes large values when the correlogram takes small values and vice versa, so that the ridges now show up as darker, rather than lighter, shades of grey. It was precisely to emphasize this relationship that we scaled the variogram; the scaling does not affect the shape of the plot, and is usually omitted.

So why bother with the variogram? For analysing lattice data, it adds nothing to the spatial correlogram. But for irregularly spaced data such as the Galicia lead pollution data, it is not obvious how we would begin to calculate a spatial correlogram, whereas we can easily calculate the squared differences between pairs of measurements and investigate how these do or do not relate to spatial separation.

10.5 A case-study in spatial prediction: mapping lead pollution in Galicia

10.5.1 *Galicia lead pollution data*

We now use the data represented schematically in Figure 4.20 to show how the ideas in Section 10.4 can be extended to deal with an irregular spatial distribution of sampling locations, and how the results can then be used to produce a pollution map for the whole of Galicia.

10.5.2 *Calculating the variogram*

To handle spatial data collected at an arbitrary set of locations, it is helpful first to establish a little more notation.

Each data point constitutes a set of three numbers, (x, y, z), where x and y define a location and z is the corresponding measured value. We write the complete set of data as $(x_i, y_i, z_i) : i = 1, \ldots, n$. For the Galicia lead pollution data, $n = 63$. Now, for any pair of data-points i and j, define $dx_{ij} = x_i - x_j$, $dy_{ij} = y_i - y_j$ and $v_{ij} = \frac{1}{2}(z_i - z_j)^2$. A three-dimensional plot of the points $(dx_{ij}, dy_{ij}, v_{ij})$ is called a *variogram cloud*. Figure 10.8 illustrates this for an artificially small synthetic dataset consisting of $n = 16$ measurements. Note that for any pair i and j, $v_{ji} = v_{ij}$, so the figure necessarily shows mirror symmetry about each axis. The right-hand panel of Figure 10.8 shows a clear directional effect, with the larger values of v_{ij} concentrated around the southwest to northeast axis and, along this axis, a distance effect with larger values of v_{ij} towards the two corners. For realistically large datasets, the variogram cloud becomes very cluttered and difficult to interpret. The usual practice is then to 'bin' the cloud by dividing the (x, y)-space into square cells, or bins, and to calculate the average of the v_{ij} corresponding to the pairs (dx, dy) falling into each bin.

In our synthetic example, a possible set of bins is indicated by the dashed grid lines.

When the binned variogram cloud does not show evidence of directional effects, its definition and the associated calculations can be simplified by reducing each pair (dx_{ij}, dy_{ij}) to a distance, $u_{ij} = \sqrt{(dx_{ij}^2 + dy_{ij}^2)}$. Then, the cloud becomes a two-dimensional scatterplot of the points (u_{ij}, v_{ij}) and binning reduces to averaging the v_{ij} corresponding to each of a set of distance intervals. Any directional effects are necessarily hidden by the averaging of results from all pairs of measurements separated by the same distance, irrespective of their orientation. However, when this averaging is appropriate the result is usually easier to interpret. Figure 10.7 shows an example, using the same synthetic data as for Figure 10.8.

Figure 10.9 shows the binned directional variogram of the Galicia lead pollution data. Note that because of the variogram's inherent mirror symmetry, Figure 10.9 covers only positive lags v but both positive and negative lags u. The diagram shows no obvious directional effect of the kind that we saw in the wheat uniformity trial data.

Figure 10.10 shows the isotropic variogram of the Galicia lead pollution data. This reveals a clear distance effect whereby the variogram increases, i.e., spatial correlation becomes weaker, with increasing separation distance.

10.5.3 *Mapping the Galicia lead pollution data*

We now consider how the behaviour of the variogram, as estimated by Figure 10.10, can suggest how we might best use the data to construct a map of the underlying pollution surface.

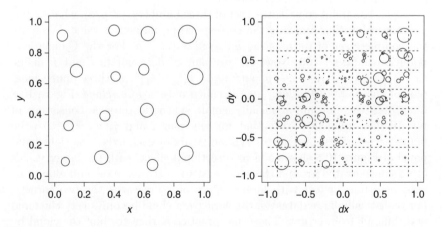

Fig. 10.7. A synthetic set of $n = 16$ measurements and their directional variogram cloud.

A CASE-STUDY IN SPATIAL PREDICTION

149

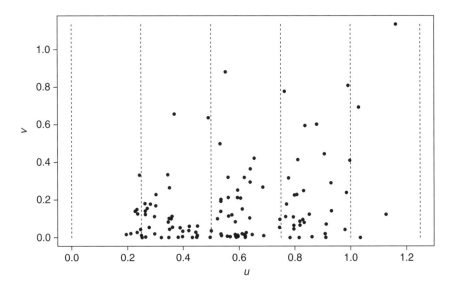

Fig. 10.8. A synthetic set of $n = 16$ measurements and their isotropic variogram cloud.

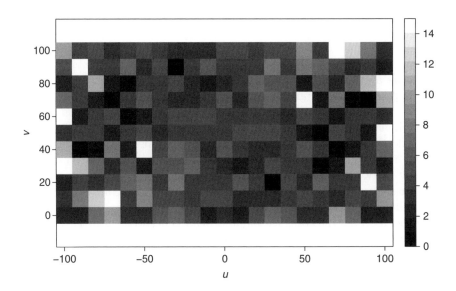

Fig. 10.9. Directional variogram of the Galicia lead pollution data.

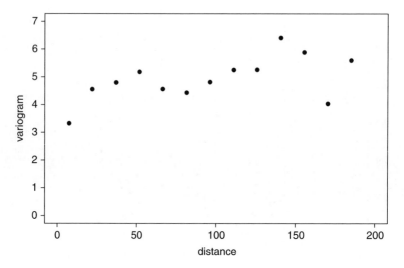

Fig. 10.10. Isotropic variogram of the Galicia lead pollution data.

Firstly, what value would we predict for the lead concentration *at* one of the $n = 63$ sampled locations? At first sight, the answer seems obvious – haven't we already measured this quantity? But remember that we measured the lead concentration in a *sample* of moss gathered at the location in question. Would a second sample have given the same answer? Presumably not. But if it would have given a different answer, then how different? The variogram gives us a clue. Remember that the plotted points in Figure 10.10 are averages of (half) squared differences between lead concentrations at (approximately) a given separation distance. As already noted, the overall impression conveyed by Figure 10.10 is that this average is a fairly smooth curve that increases with increasing distance, and conversely. If we imagine this curve being extrapolated back towards zero distance, where do we think it would cut the y-axis? Figure 10.11 shows what you might accept as two reasonable extrapolations. One of these would imply that the variogram approaches zero as the separation distance approaches zero, the other would not. If we prefer the former extrapolation, we would have to conclude that the observed value of lead concentration at each of the 63 sampled locations gives the best possible prediction of the underlying concentration at that location – indeed, it *is* the underlying concentration. If, on the other hand, we prefer the second extrapolation, then we might do better by predicting the underlying concentration at a sampled location as a weighted average of measured values at nearby sample locations. The justification for this lies in the fact that the variogram increases in value with increasing separation distance. This implies that the values from close locations are similar, and hence that an average of several might give a

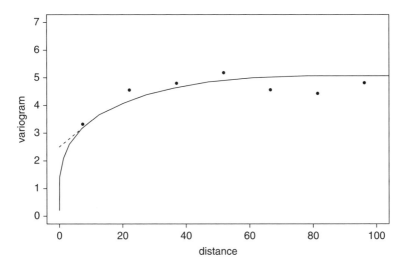

Fig. 10.11. Isotropic variogram of the Galicia lead pollution data with two fitted curves extrapolated to zero distance.

more precise prediction, i.e., one with smaller variance, without introducing much bias. In the authors' opinion, resolving this ambiguity usually needs more than empirical statistical evidence – it needs scientific knowledge. Note, for example, that in the specific context of the Galicia lead pollution data, the reproducibility of measured concentrations at any one location depends on two considerations: how precise is the assay that extracts the lead measurement from a sample of moss? and how different are the true lead concentrations likely to be in different samples from the same location?

Turning the above heuristic argument into a specific set of 'best' predictions is not straightforward, and the technical details are beyond the scope of this book. But it turns out that a general method of spatial prediction, with good statistical properties, can be represented by the formula

$$\hat{S}(x) = \sum_{i=1}^{n} w_i(x) Y_i, \tag{10.1}$$

where the Y_i are the measured values and the weights, $w_i(x)$, can be calculated from the variogram of the data. Typically, the $w_i(x)$ will be large when the sampling location x_i is close to the prediction location x, and conversely. How large, and how close, depend on the exact form of the variogram.

The formal status of equation (10.1) is that it defines, for any location x, the mean of the distribution of the unknown value of $S(x)$ given all of the information provided by the data, under a model that assumes firstly that $S(x)$ follows a Normal distribution, and secondly that the correlation between values of $S(x)$ at any two different locations can be described by a function of the distance between those two locations. We call this distribution the *predictive distribution* of $S(x)$. In practice, equation (10.1) can still give good results when these assumptions do not hold, but it is most likely to do so when the measurements Y_i are distributed more or less symmetrically about their mean. When this is not the case, it is worth considering whether to transform the data before fitting a theoretical model to the variogram.

Figure 10.12 shows histograms of the 63 observed values of lead concentrations and of log-transformed lead concentrations. The log-transformation produces a somewhat more symmetric distribution. Another, and arguably better, reason for favouring the log-transformation is that changes in environmental policy would be more likely to engender relative, rather than absolute, changes in pollution levels. For the formal modelling of these data we therefore define the response, Y_i, to be the log-transformed lead concentration.

Figure 10.13 shows the estimated variogram of the log-transformed data, together with a smooth curve that is the theoretical variogram of a parametric model. The general form of this model is

$$V(u) = \tau^2 + \sigma^2\{1 - \exp(-u/\phi)\} : u \geq 0. \tag{10.2}$$

The curve in Figure 10.13 uses the maximum likelihood estimates of the three parameters, namely $\tau^2 = 0.083$, $\sigma^2 = 0.1465$ and $\phi = 19.3045$. It is

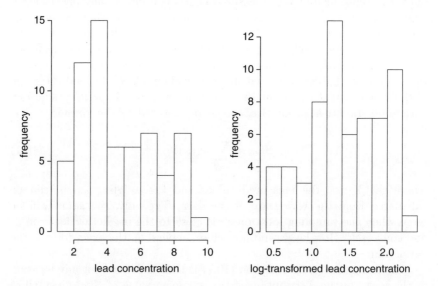

Fig. 10.12. Histograms of the 63 measured values of lead concentration (left-hand panel) and log-transformed lead concentration (right-hand panel) from the Galicia biomonitoring data.

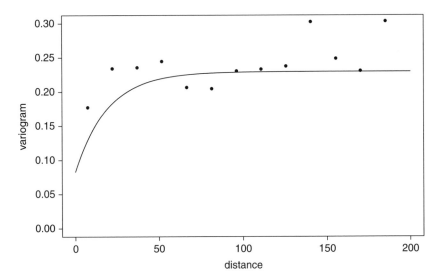

Fig. 10.13. Isotropic variogram (solid dots) of the lead pollution data and theoretical model (smooth curve) estimated by maximum likelihood.

worth emphasizing that the fitted line in Figure 10.13 is *not* the conventionally defined 'line of best fit' to the empirical variogram. As discussed in Chapter 3, the best fit of any statistical model to a set of data is obtained by using the likelihood function rather than by ad hoc methods. For the linear regression models that we described in Sections 7.5 and 7.6, the line of best fit *is* the maximum likelihood estimate; for the geostatistical model that we are using for the Galicia lead pollution data, it emphatically is not.

We now use the theoretical model (10.2) in conjunction with the maximum likelihood estimates of the model parameters to calculate the weighting functions $w_i(x)$ in (10.1) and so obtain a predicted value, $\hat{S}(x)$, for any location x. This gives the predictive map shown in Figure 10.14. Notice that the map captures the general spatial distribution of lead concentrations that were apparent in the original data in Figure 4.20, but additionally gives objectively determined predictions at all unsampled locations.

Figure 10.14 gives no indication of how precisely or imprecisely we have been able to predict the underlying lead concentration surface. Each mapped value is the mean of the predictive distribution of $S(x)$. In Chapter 6, we emphasized the importance of quoting not just an estimated mean but its associated confidence interval. We could use the simple 'mean plus and minus two standard errors' rule to construct analogues of confidence intervals for each of the mapped values, but because of the complexity of a spatial surface by comparison with a single number, a better strategy is to

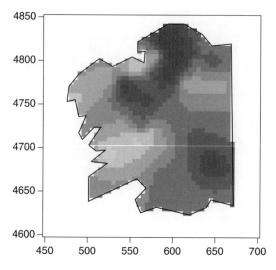

Fig. 10.14. Mean predicted pollution surface for the Galicia lead pollution data. The colour-scale runs from 2.0 (red) through yellow to 10.0 (white). Mapped values range from 2.8 to 7.8. This figure is reproduced in colour in the colour plate section.

simulate a large number of surfaces from the joint predictive distribution of $S(x)$ for all locations x of interest, and use these to construct confidence intervals for whatever properties of the complete surface are of interest. Figure 10.15 shows three maps, corresponding to the lower quartile, median and upper quartile of the predictive distribution at each location. Not untypically, the median map (central panel of Figure 10.15) is similar to the mean map shown in Figure 10.14 and either serves equally well as a set of point predictions. Mapping lower and upper quartiles is analogous to using a 50% confidence interval. Even with this low (by conventional standards) level of coverage, the difference between the lower and upper quartile maps is substantial: attempting to map a complex spatial surface from only 63 values is an ambitious task.

A policy maker might be interested in how precisely we could estimate the highest levels of pollution Galicia-wide. Reading off the maximum value of the surface shown in Figure 10.14 would not be a sensible way to answer this question. The maximum value on the map is approximately 8.0, but the data include a value of 9.5 which, even allowing for measurement error in the data, tells us that 8.0 must be an underestimate. Instead, we again use the method of simulating from the predictive distribution. From each simulation, we record the maximum predicted value, and by running repeated simulations build up a sample of values from the predictive distribution of the Galicia-wide maximum lead concentration. Figure 10.16 shows the resulting histogram, with vertical dashed lines defining a range containing

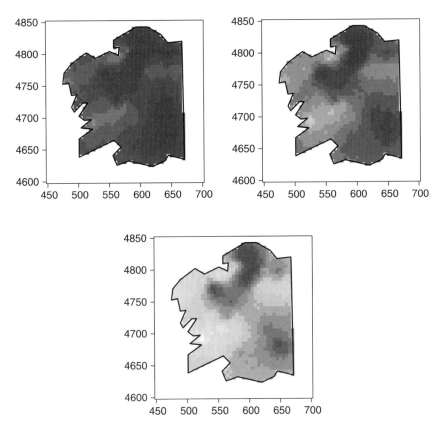

Fig. 10.15. Lower quartile, median and upper quartile predicted pollution surfaces for the Galicia lead pollution data. The colour-scale runs from 2.0 (red) through yellow to 10.0 (white). This figure is reproduced in colour in the colour plate section.

95% of the values. This range is called a 95% prediction interval, and is analogous to a 95% confidence interval for a model parameter. Note again that the limits of uncertainty are wide, but better an honest prediction than a dishonestly precise one.

The method of spatial prediction described in this section is called 'ordinary kriging' (see below for an explanation of the name). We make no claim that this is always the best method to use. But it does have several advantages over other, more ad hoc, methods. Firstly, when used in conjunction with maximum likelihood estimation of the variogram parameters, it is entirely objective. Secondly, and provided the underlying theoretical model is correct, it can be shown to be the best possible method in the specific sense of making the average squared error, $\{\hat{S}(x) - S(x)\}^2$, as small as possible. The proviso of a correct theoretical model is important, but

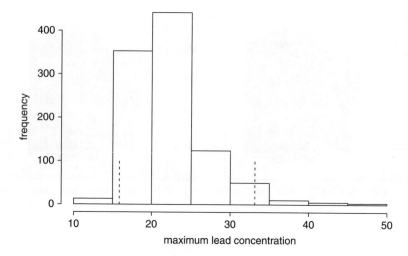

Fig. 10.16. Historgram of a sample of values from the predictive distribution of the Galicia-wide maximum lead concentration. Vertical dashed lines define a range containing 95% of the sampled values.

should not be interpreted literally. We can use all of the well-established techniques of statistical modelling, including but not restricted to those described in this book, to check whether the assumed model fits the data well. If so, then even if not strictly correct it can be taken to be a reasonable approximation to the truth, and used to answer the scientific questions posed by the data in hand.

10.6 Further reading

The branch of spatial statistics known as geostatistics has its origins in the South African mining industry, where Professor D. G. Krige advocated the use of statistical methods to assist in the practical problem of predicting the likely yield that would be obtained from mining at a particular region under consideration for possible exploitation; see, for example, Krige (1951). Geostatistics is sometimes described as a self-contained methodology with its own terminology and philosophy of inference that are somewhat removed from the statistical mainstream; see, for example, Chilès and Delfiner (1999). Books that describe geostatistical methods within a mainstream statistical setting include Cressie (1991) and Diggle and Ribeiro (2007). An easier read than any of these is Waller and Gotway (2004, Chapter 8).

In this chapter, we have concentrated on geostatistical methods, and have touched briefly on the analysis of lattice data. A third major branch of spatial statistics is *spatial point processes*. This term refers to natural processes that generate data in the form of a set of locations. Examples include the locations of individual trees in a forest, of individual cases of a disease in a human or animal population, or of cell nuclei in a section of biological tissue. Methods for analysing spatial point process data are described in Diggle (2003), in Ilian, Penttinen, Stoyan and Stoyan (2008) and in Waller and Gotway (2004, Chapter 5).

Spatial statistical methods involve technically sophisticated ideas from the theory of stochastic processes, and computationally intensive methods of data analysis. Our excuses for including them in this introductory book, apart from personal bias, are two-fold. Modern advances in data-collection technology, including microscopy, remotely sensed imaging and low-cost GPS devices, are all making the availability of spatial data increasingly common in many natural, biomedical and social scientific disciplines; allied to this, spatial statistical methods are beginning to be incorporated into geographical information systems. This is liberating for users who are not formally trained statisticians, but at the same time potentially harmful if it leads to unthinking analyses using data that are incompatible with the implicit modelling assumptions that the software invokes.

Appendix
The R computing environment

A.1 Background material

R is an open-source product, built on the S language (Becker, Chambers and Wilks, 1988) but with added statistical functionality provided by a dedicated project team and a vast array of contributed packages. It is now the software of choice for the international statistical research community and is easily installed and run within most common operating systems, including Windows, Linux and Mac.

Many instructional books on R have now been written, either covering the basic system or linked to particular statistical applications. Examples in the first category that we have found useful and accessible include Venables and Smith (2002) or Dalgaard (2002). In the second category, probably the best-known example is the classic text by Venables and Ripley (2002), which covers many statistical topics at a fairly advanced level. A better choice for beginners is Verzani (2004).

A straightforward R session consists of a series of commands whose generic syntax echoes classical mathematical notation, as in y=f(x), meaning that the R function f(.) processes the argument x to give the result y. Functions can be almost arbitrarily complex, with many arguments, each of which can be, amongst other things, a single number, a vector, a matrix, a character string or combinations of the above combined into a list. Similarly, the result can be any of the above, and can include graphical and/or tabular output.

Functions can be of three kinds: in-built; user-written; or added-on as components of a package. In Section A.3 we show an example of an R session that illustrates the first two of these. A key feature of R is that the results of a function become additional objects that can be manipulated in their own right; for example, they can be displayed on the screen, written to a file for safe keeping or used as arguments to other functions. Hence, the 'equals' sign in y=f(x) should be seen as an assignment of information rather than a logical equality. Some users (amongst whom we count ourselves) prefer to use the equivalent syntax y<-f(x) to emphasize this, but we are a dying breed.

A.2 Installing R

The R software, and contributed packages, can be accessed and installed by visiting the R Project website, www.r-project.org, and navigating through the various options. Here, we give step-by-step instructions for computers running a Windows operating system. The differences for Mac or Linux operating systems are minor.

To download the software, proceed as follows (notes are in parentheses).

(1) Go to http://www.r-project.org/
(2) Click on CRAN
(3) Click on the web-address for one of the 'CRAN Mirrors'
    (The choice is not critical, but you are recommended to choose a mirror from a nearby country.)
(4) Inside the box 'Download and install R' click on Windows
    (This downloads the binary version of the software – if you want the source code, you are probably already enough of a software expert not to need these step-by-step instructions.)

Many of the statistical methods that you are likely to want to use are now available to you. If they are not, they may well be available amongst the wide variety of add-on *packages* that have been written and documented by statisticians world-wide. For example, the package splancs offers additional functionality to analyse spatial point process data.

Depending on exactly how your computer is connected to the Web, you may be able to install packages within an R session using the install.packages() function; an example of an R session follows in Section A.3. If this does not work, you can proceed as follows.

(1) Go to http://www.r-project.org/
(2) Click on CRAN
(3) Click on the web-address for one of the 'CRAN Mirrors'
(4) Click on Packages
    (Under the heading 'Available packages' you can now scroll to the package you want. At the time of writing, there were 2428 packages available so you probably want to know what you are looking for before you start – but see *Task Views* below)
(5) Click on the name of the package you want
    (This takes you to a page where you can download the package and assocated documentation.)

Once a package has been installed, its functionality and associated documentation are accessible in exactly the same way as for built-in R functions. The R code available from the book's web page includes several examples.

If you don't know exactly what package you need, you should find *Task Views* helpful. These are guides to what packages are available under a number of broad headings (26 at the time of writing, somewhat more manageable than 2428). To access Task Views, proceed as follows:

(1) Go to http://www.r-project.org/
(2) Click on CRAN
(3) Click on the web-address for one of the 'CRAN Mirrors'
(4) Click on Task Views

A.3 An example of an R session

To open an R session, you will need to activate the software, either by clicking on its icon or, if you are working on a command-line system, typing the command R. Within the R software itself, you need to supply a sequence of commands, each one in response to the prompt >. You can do this interactively, or you can prepare a sequence of R commands in a plain text file and run them either by cutting-and-pasting or more elegantly, by instructing R to look in the file that contains your sequence of commands. In our experience, a combination of these strategies works best. At an exploratory stage, we tend to use R in interactive mode, whereas when conducting a formal analysis we type and store sequences of commands into text files to keep a permanent record of what we have done, the book's web page being a case in point.

In the following, lines beginning with the # symbol are comment lines: their only purpose is to explain what the software is doing. Other lines are command lines: each of these should generate a response from the software if you enter them into an R window.

```
# assign the integers 1 to 100 to a vector, with name x, and
# print the first 5 values onto the screen
x<-1:100
x[1:5]
# simulate data from a quadratic regression model whose
# residuals are Normally distributed, standard deviation 12
mu<-2+0.5*x+0.01*x*x
z<-12*rnorm(100)
y<-mu+z
# display the first five (x,y) pairs (no assignment), with
# y-values rounded to 3 decimal places
cbind(x[1:5],round(y[1:5],3))
# draw a scatterplot of x against y
```

```
plot(x,y)
# customize the scatterplot using optional arguments to the
# plot() function
plot(x,y,pch=19,col="red",cex=0.5,main="plot of x vs y")
# fit a linear regression model to the simulated data
# and summarize the result
fit1<-lm(y~x)
summary(fit1)
# list the names of the components of the R object that
# stores information about the fitted model
names(fit1)
# these components are individually accessible using the
# $ sign to indicate which component you want
fit1$coef
alpha<-fit1$coef[1]
beta<-fit1$coef[2]
# add the fitted regression line to the scatterplot
lines(x,alpha+beta*x)
# the plot now shows clearly that the linear model does not
# give good fit to the data, so we now fit the
# quadratic model and add the fitted quadratic curve
xsq<-x*x
fit2<-lm(y~x+xsq)
alpha<-fit2$coef[1]
beta<-fit2$coef[2]
gamma<-fit2$coef[3]
muhat<-alpha+beta*x+gamma*xsq
lines(x,muhat,col="blue",lwd=2)
# we'll now write our own function to fit and plot the
# quadratic regression model
quadfit<-function(x,y,plot.result=F,x.plot=x) {
# Arguments:
#   x: values of the explanatory variable
#   y: values of the response variable
#   plot.result: if T (true), plot of data and fitted
#                regression curve is produced,
#                if omitted plot is not produced because
#                default is F (false)
#   x.plot: values of the explanatory variable used to draw
#           the fitted curve, if omitted uses same values as
#           x, but sorted into ascending order
# Result:
#   A standard R model object, plus (if plot.result=T) a plot
#   of data and fitted regression curve drawn on the current
```

```
#    plotting device
   xsq<-x*x
   fit<-lm(y~x+xsq)
   alpha<-fit$coef[1]; beta<-fit$coef[2]; gamma<-fit$coef[3]
   if (plot.result==T){
      plot.x<-sort(plot.x)
      plot(x,y,cex=0.5, col="red",main= "data and quadratic fit")
      y.plot=alpha+beta*x.plot+gamma*x.plot*x.plot
      lines(x.plot,y.plot,lwd=2,col= "blue")
      }
   fit
   }
# now use this function in default and non-default modes
quadfit(x,y)
# this writes the output to the screen but does not save it
# for future use, nor does it produce a plot
fit3<-quadfit(x,y,plot.result=T,x.plot=0.2*(0:500))
# the plotting option is now invoked by typing 'plot.result=T'
# (T for true), also typing 'plot.x=0.2*(0:500))' gives a
# smoother curve than the default would have done,
# other output is assigned rather than displayed - but can be
# displayed if you wish, note that fit2 and fit3 should be
# identical - check by displaying summaries
summary(fit2)
summary(fit3)
# now quit R - you will be invited to save your work if you
# wish, in which case it will be loaded when you next run the
# R program
q()
```

References

Altman, D. G. (1991). *Practical Statistics for Medical Research*. London: Chapman and Hall.

Anderson, A. J. B. (1989). *Interpreting Data*. London: Chapman and Hall.

Armitage, P., Bodmer, W., Chalmers, I., Doll, R. and Marks, H. (2003). Contributions to a symposium on Bradford Hill and Fisher. *International Journal of Epidemiology*, **32**, 922–948.

Becker, R. A., Chambers, J. M. and Wilks, A. R. (1988). *The new S language*. Pacific Grove, CA: Wadsworth and Brooks.

Box, J. F. (1978). *R. A. Fisher: The Life of a Scientist*. New York: Wiley.

Chatfield, C. (1996). *The Analysis of Time Series: An Introduction. 5th edition*. London: Chapman and Hall.

Chetwynd, A. G. and Diggle, P. J. (1995). *Discrete Mathematics*. Oxford: Butterworth-Heinemann.

Chilès, J.-P. and Delfiner, P. (1999). *Geostatistics*. New York: Wiley.

Cleveland, W. S. (1984). *Elements of Graphing Data*. Belmont, CA: Wadsworth.

Cleveland, W. S. and McGill, M. E. (1988). *Dynamic Graphics for Statistics*. Belmont, CA: Wadsworth.

Collett, D. (2003). *Modelling Survival Data in Medical Research*. New York: Chapman and Hall.

Cox, D. R. (1972). Regression models and life-tables (with Discussion). *Journal of the Royal Statistical Society,* B **34**, 187–220.

Cox, D. R. and Oakes, D. (1984). *Analysis of Survival Data*. New York: Chapman and Hall.

Cressie, N. A. C. (1991). *Statistics for Spatial Data*. New York: Wiley.

Dalgaard, P. (2002). *Introductory Statistics with R*. New York: Springer.

Diggle, P. J. (1990). *Time Series*. Oxford: Oxford University Press.

Diggle, P. J. (2003). *Statistical Analysis of Spatial Point Patterns. 2nd edition*. London: Edward Arnold.

Diggle, P. J., Heagerty, P., Liang, K. Y. and Zeger, S. L. (2002). *Analysis of Longitudinal Data. 2nd edition*. Oxford: Oxford University Press.

Diggle, P. J. and Ribeiro, P. J. (2007). *Model-based Geostatistics*. New York: Springer.

Dobson, A. (2001) *An Introduction to Generalized Linear Models. 3rd edition*. Boca Raton, FL: Chapman and Hall.

Fanshawe, T. R., Diggle, P. J., Rushton, S., Sanderson, R., Lurz, P. W. W., Glinianaia, S. V., Pearce, M. S., Parker, L., Charlton, M. and Pless-Mulloli, T. (2008). Modelling spatio-temporal variation in exposure to particulate matter: a two-stage approach. *Environmetrics*, **19**, 549–566.

Fernández, J. A., Rey, A. and Carballeira, A. (2000). An extended study of heavy metal deposition in Galicia (NW Spain) based on moss analysis. *Science of the Total Environment*, **254**, 31–44.

Fitzmaurice, G. M., Laird, N. M. and Ware, J. H. (2004). *Applied Longitudinal Analysis*. Hoboken, NJ: Wiley.

Gardner, M. J. and Altman, D. G. (1986). Confidence intervals rather than P values: estimation rather than hypothesis testing. *British Medical Journal*, **292**, 746–750.

Gleick, J. (2003). *Isaac Newton*. New York: Alfred A. Knopf.

Graff-Lonnevig, V. and Browaldh, L. (1990). Twelve hours bronchodilating effect of inhaled Formoterol in children with asthma: a double-blind cross-over study versus Salbutamol. *Clinical and Experimental Allergy*, **20**, 429–432.

Hall, A. R. (1980). *Philosophers at War: The Quarrel Between Newton and Gottfried Leibniz*. Cambridge: Cambridge University Press.

Hempel, S. (2006). *The Medical Detective: John Snow and the Mystery of Cholera*. London: Granta.

Ilian, J., Penttinen, A., Stoyan, H. and Stoyan, D. (2008). *Statistical Analysis and Modelling of Spatial Point Patterns*. Chichester, UK: Wiley.

Iliffe, R. (2007). *Newton. A Very Short Introduction*. Oxford: Oxford University Press.

Kalbfleisch, J. D. and Prentice, R. L. (2002). *The Statistical Analysis of Failure Time Data*. New York: Wiley.

Kalman, R. E. (1960). A new approach to linear filtering and prediction problems. *Journal of Basic Engineering*, **82**, 35–45.

Kalman, R. E. and Bucy, R. S. (1961). New results in linear filtering and prediction problems. *Journal of Basic Engineering*, **83**, 95–108.

Kaplan, E. L. and Meier, P. (1958). Nonparametric estimation from incomplete observations. *Journal of the American Statistical Association,* **53**, 457–481.

Krige, D. G. (1951). A statistical approach to some basic mine valuation problems on the Witwatersrand. *Journal of the Chemical, Metallurgical and Mining Society of South Africa*, **52**, 119–39.

Lawless, J. F. (2003). *Statistical Models and Methods for Lifetime Data. 2nd edition.* Hoboken, NJ: Wiley.

Lindley, D. V. (2006). *Understanding Uncertainty*. Hoboken, NJ: Wiley.

McBratney, A. B. and Webster, R. (1981). Detection of ridge and furrow pattern by spectral analysis of crop yield. *International Statistical Review*, **49**, 45–52.

McColl, J. H. (1995). *Probability*. Oxford: Butterworth-Heinemann.

McCullagh, P. and Nelder, J. A. (1989). *Generalized Linear Models. 2nd edition*. London: Chapman and Hall.

Medical Research Council (1948). Streptomycin in tuberculosis trials committee. Streptomycin treatment of pulmonary tuberculosis. *British Medical Journal*, **2**, 769–783.

Mercer, W. B. and Hall, A. D. (1911). The experimental error of field trials. *Journal of Agricultural Science*, **4**, 107–132.

Meinhold, R. J. and Singpurwalla, N. D. (1983). Understanding the Kalman filter. *The American Statistician*, **37**, 123–127.

Nelder, J. and Wedderburn, R. (1972). Generalized linear models. *Journal of the Royal Statistical Society,* A **135**, 370–384.

Piantadosi, S. (2005). *Clinical Trials: A Methodological Perspective.* Hoboken, NJ: Wiley.

Popper, K. (1959). *The Logic of Scientific Discovery.* New York: Basic Books.

Senn, S. J. (2002). *Cross-over Trials in Clinical Research. 2nd edition.* Chichester, UK: Wiley.

Tufte, E. R. (1983). *The Visual Display of Quantitative Information.* Cheshire, CT: Graphics Press.

Venables, W. N. and Ripley, B. D. (2002). *Modern Applied Statistics with S. 4th edition.* New York: Springer.

Venables, W. N. and Smith, D. M. (2009). *An Introduction to R. 2nd edition.* Bristol: Network Theory Limited.

Verzani, J. (2004). *Using R for Introductory Statistics.* Boca Raton, FL: Chapman and Hall/CRC.

Wainer, H. (1997). *Visual Revelations.* New York: Springer.

Wainer, H. (2005). *Graphic Discovery.* Princeton: Princeton University Press.

Waller, L. and Gotway, C. A. (2004). *Applied Spatial Statistics for Public Health Data.* New York: Wiley.

Wilson, D. J., Gabriel, E., Leatherbarrow, A. J. H., Cheesbrough, J., Gee, S., Bolton, E., Fox, A., Fearnhead, P., Hart, C. A. and Diggle, P. J. (2008). Tracing the source of campylobacteriosis. *Public Library of Science Genetics,* **4**, e1000203 (doi:10.1371/journal.pgen.1000203).

Young, P. and Beven, K. J. (1994). Data-based mechanistic modelling and the rainfall-flow nonlinearity. *Environmetrics,* **5**, 335–363.

Index

Note: page numbers in *italics* refer to Figures and Tables.

2 × 2 factorial experimental design 33, 65
5% critical value 29

Abram, Bev, *Arabadopsis* investigation 33–4
accelerated life model 126
additive effects 66
age effects, incorporation into proportional hazards model 125
agricultural field experiments 57–9
　blocking 63–5
　factorial experiments 65–7
　randomization 59–63
　variation in yields, summary measure 64
air pollution 141, *142*
　biomonitoring 141, 143
　black smoke monitoring stations, Newcastle upon Tyne *142*
air quality measurement 141
alternative hypothesis 68
animation, use in data displays 51
Anscombe's quartet 99
　residual against fitted value plots *100*
Arabadopsis investigation 33–4
　comparing multiple batches of data 40–2
　data *35*
　data analysis 92–5
　interactions between factors 65–7
　relationships between variables 42–5
　single batch data display 36–40
assumptions checking 97–101
　residual diagnostics 99–101, *102*
asthma 71
asthma trial
　data *72*
　data analysis 73–5, 77
　general linear model 90–2
autocorrelation 130–1
　correlograms 131–*3*
　use in prediction 134, 135–7

axiomatic approach to probability 31

bar-charts 38, *39*
baseline hazard function 123
Bayes, Thomas 32
Bayesian inference 31–2
binary data
　logistic model 108–9
　odds ratio (OR) 55–6
black smoke monitoring stations, Newcastle upon Tyne *142*
blinding 67
blocking 63–5
　paired design 73
　randomized block design 78
blood pressure measurements
　relationship to BMI 18–20
　variation between patients 18

calculus, discovery 6
Campylobacter jejuni, genetic profiles study 49, *50*
categorical data, graphical presentation 49
cDNA (red-green) microarrays 33, *34*
censored data 115
cholera, London epidemic, 1854 48
climate, distinction from weather 127
clinical significance 69–70
clinical trials 67–8
　crossover designs 77–8
　drug treatments for asthma 71
　hypothesis testing 68–9
　paired design 73
　　data analysis 73–5
　parallel group design 72–3
　　data analysis 75–6
　power calculations 69–70
　primary outcome 68

clinically useful differences 27
colour, use in data displays 51
comparative experiments 71–2
    comparing more than two treatments 78, 92–5
    crossover designs 77–8
    paired design 73
        data analysis 73–5
    parallel group design 72–3
        data analysis 75–6
completely randomized designs 61–3, 78
    block design 64
computation ix
    see also R software
confidence intervals 26–7
    paired experimental design 74–5
    parallel group design 75–6
confirmatory data analysis 9
confounders 8, 125
constants, notational conventions 12
correlation 54–5
correlogram 131–*3*
    spatial 144–6
count data
    error distribution 112
    log-linear model 110–1
covariates (explanatory variables) 18–20
crossover trials 77–8
cumulative plots *37*

data analysis, exploratory *see* exploratory data analysis
data-based mechanistic modelling 16
data exploration 9–10
data modelling 10–12, 15–16
deductive method vii, 24
degrees of freedom 93
dependence, repeated measurements 22–3
design variables 17
    difference from explanatory variables 20
deterministic variation 127, 128
diagnostic checking 98
    residual diagnostics 98–101, *102*
dialysis 114
dialysis data
    Kaplan–Meier estimates 118–*19*
    proportional hazards model 123–5
disease prevalence estimation 27–30
dot-plots *36–7*
Drew, Peter, peritoneal dialysis data 114, *115*

dropouts 47
drug comparisons 18–20

efficiency of experiments 8
empirical models 11, 83
equipoise, clinical trials 67
error distributions 112
ethical considerations, clinical trials 67
experimental design 8–9, 15
    agricultural field experiments 57–9
    blocking 63–5
    clinical trials 67–8
    comparative experiments 71–3
    factorial experiments 65–7
    observational studies 70
    randomization 59–63
    statistical significance and statistical power 68–70
experimental error 81, 82
explanatory variables (covariates) 18–20, multiple 90
exploratory data analysis 9–10
    comparing multiple batches of data 40–2
    displaying relationships between variables 42–5
    displaying single batches of data 36–40
    goals 35–6
    graphical design 50–1
    numerical summaries
        summarizing relationships 54–6
        summarizing single and multiple batches of data 53–4
        summation notation 51–2
    of proportions 49–50
    of spatial data 48–9
    of time series 45–7
exponential growth model 103–6
    comparison with logistic growth model *107*
extraneous variation 8

factorial experiments 65–7
factors 17, 79, 90
    in general linear model 90
falsification of theories 2
Fisher, R.A. 59, *60*
fitted values 98
five-number summary 53–4
frequency polygons *38*

Galicia lead pollution data 143
    mapping 150–6

INDEX

pollution surfaces *154*, 155
  variogram calculation 147–8, *149*, *150*
gene expression microarrays 33–5
  comparing multiple batches of data 40–2
  displaying relationships between variables 42–5
  displaying single batches of data 36–40
general linear model 90
  assumptions 97
  comparing more than two treatments 92–5
  inclusion of factors 90
generalized linear models 108–12
  fitting to data sets 111–2
  logistic model for binary data 108–9
  log-linear model for count data 110–1
geography, first law of 64, 144
geostatistical datasets 48–9, *50*
geostatistics, further reading 156
glyphosphate data *79*–*80*
  fitting a model 96–7
  residual diagnostics 101, *102*
  transformation *87*–8, *89*
graphical design 50–1

haemo-dialysis 114
hazard functions 120–2
hazard ratio 124
Hill, Austin Bradford 67
histograms 38–*9*
  comparing multiple batches of data 40, *41*, 42
hypothesis testing 14–15, 26, 68–70
  likelihood ratio testing 29–30

incomplete randomized block designs 64
independence of residuals 101
independent events 21–2
inference vii, 13, 16, 24
  Bayesian 31–2
  confidence levels 26–7
  estimates 25–6
  hypothesis testing 14–15
  parameter estimation 13–14
  populations 25
  prediction 15
  samples 25
informed consent 20, 67–8
input variables 10, 11, 57
interactions between factors 65–6
inter-quartile range 53, 54
interval-censored data 115

interval estimates 14, 25–6
intrinsically non-linear models 107
inverse logit function 109, *109*, *110*

Kalman filter 139–40
Kaplan–Meier survival curves 117–19, *118*
kidney failure 114
kidney failure data *115*
  Kaplan–Meier estimates 118–*19*
  proportional hazards model 123–5
Kolmogorov, Andrey 31
Krige, D.G. 156
kriging 155

lag-*k* autocorrelation 130
lattice data, spatial correlogram 144, *145*
least squares estimation 87
  weighted least squares criterion 88
left-censored data 115
Leibnitz, Gottfried Wilhelm von 6
levelled scatterplots 44–*5*
life expectancy 119
  hazard functions 120–2
lifetime distributions *120*
  comparison with hazard functions *121*–*2*
likelihood-based estimation 95
  glyphosphate data 95–7
likelihood 27–31, 95
  likelihood function 27–31, *29*
    maximum likelihood estimates 28–9, 95
  likelihood ratio testing 29–30
linear predictor 111
linear relationships 11–12,
line of best fit 86 *86*
link function 112
log-likelihood function 28, *29*
logistic growth model 107, *107*
logistic-linear regression model (logistic model) 108–9
longitudinal data 47

main effect 65–6
marked spatial point pattern 48, *49*
mathematical model 81
  see also modelling
mean 52
mean square error (MSE) 137–*8*
mechanistic model 10–11, 83
median 53, 54

modelling 10–12, 15–16, 79–80
　exponential growth model 103–6
　general linear model 90
　　comparing more than two
　　　treatments 92–5
　　inclusion of factors 90
　generalized linear models 108–12
　　fitting to data sets 111–2
　　logistic model for binary data 108–9
　　log-linear model for count data 110–1
　mechanistic and empirical models 83
　Newton's law experiment 81
　non-linear models 107–8
　simple linear model 83–5
　　fitting to data sets 86–7
modelling cycle 112–13
multiple batches of data, exploratory
　　analysis 40–2

Newton, Isaac 5, *6*
Newton's law, mathematical model 81
Newton's law experiment 5–6, *7*
　data *9*
　data exploration 9–10
　data modelling 10–12
　defining the question 7
　experimental design 8–9
non-linear models 107–8
non-random variables, notational
　　conventions 12
Normal distribution 84–5
notational conventions 12–13, 83–4
null hypothesis 68
numerical summaries
　summarizing relationships 54–6
　summarizing single and multiple batches
　　of data 52–4
　summation notation 51–2

observational studies 70, 125
　link between smoking and lung
　　cancer 126
odds ratio (OR) 55–6
output variables 10, 11, 57

p-values 26, 27
paired design 73
　data analysis 73–5
PANSS scores, graphical presentation 45–7
parallel group design 72–3
　data analysis 75–6
parameter estimation 13–14

parameters 13
Peak Expiratory Flow (PEF) 71
peritoneal dialysis 114
　see also dialysis data
pie charts 49, *50*
pilot studies 57
plots, agricultural field experiments 57, *58*
point estimates 13–14, 25
Poisson log-linear regression model 111
Poisson probability distributions 110 *111*
pollution *see* air pollution; smoke pollution
　time series, effect on plant growth 79
pooled sample variance 76
Popper, Karl 80
population data
　exponential growth model 103–6
　logistic growth model *107*
power calculations 58, 69–70
prediction 15, 133–7
　spatial 150–6
prediction interval 15
prediction rules, evaluation 137–8
predictive distribution 152
prescribed significance level 29, 69
probability viii, 21
　axiomatic approach 31
　further reading 31–2
　independent events 22–3
　relative frequency approach 31
probability density function (pdf) 23–4, *24*
probability distribution 23
proportional hazards model 122–3
　analysis of kidney failure data 123–5

qualitative variables 17
quantitative variables 17
quartiles 53

R software ix, 158
　example R session 160–2
　installation 159–60
random samples 25
random variables 12
random (stochastic) variation viii, 8, 17–18,
　　39
random walk 137
randomization 9, 59–63
　clinical trials 67
re-expression of laws 10
relationships between variables 42
　numerical summaries 54–6
　scatterplots 42–5

INDEX

replicates 9, 79
residual diagnostics 99–101, *102*
residual standard error 93
right-censored data 115
Rothamsted agricultural research station 57, *58*

safflower plants, effect of glyphosphate 79–*80*
sample correlation 54–5
sample mean 52
sample size calculation 69–70
sample standard deviation 52–3
sample variance 52
   pooled 76
scatterplots 9–10, *10*, *42*
   for large datasets 43–*4*
   levelled 44–*5*
   linear relationships 11–12
   use of different plotting symbols *43*
scientific laws 80
scientific method, role of statistics *2*
scientific theories 1–2
seasonality 129
selection bias 20
semi-variogram *see* variogram
simple linear model 83–5, *84*
   fitting to data sets 86–7
single batches of data, exploratory analysis 36
skewed distributions 36, 54
   Poisson probability distribution 110, *111*
smog 141, *142*
smoke pollution time series 129–*30*
Snow, John, study of London cholera outbreak 48
software ix, 4
   see also R software
spaghetti plots *47*
spatial correlogram 144, *145*, 146
spatial data
   graphical presentation 48–9
   variogram *146*–7
   for Galicia lead pollution data 147–8
spatial point patterns *48*, *49*
spatial point processes 156–7
spatial prediction, mapping Galicia lead pollution data 150–6
spatial statistics, further reading 156–7
spatial variation 144
spectral analysis 138–*9*, *140*

standard deviation 52–3
standard error 60–61, 74
statistical inference *see* inference
statistical model 82–3
   see also modelling
statistical significance 26
stochastic variation 17–18, 129
streptomycin, clinical trial 67
summary tables *53*
summation notation 51–2
survival analysis
   accelerated life model 126
   further reading 126
   proportional hazards model 122–5
survival curves 116
   Kaplan–Meier estimation 117–19, *118*
survival data (time-to-event data) 115
   regression analysis 122–3
survival functions for lifetime distributions *120*
systematic variation viii, 8, 17

t-tests 112
temperature, time series *128*
time series
   autocorrelation 130–3
   further reading 138, 140
   graphical presentation 45–7
   lack of independent replication 100–*1*, 127–8
   prediction 133–8
   seasonality 129
   spectral analysis 138–*9*, *140*
   stochastic variation 129
   trends 128–9, *130*
time-to-event data (survival data) 115
   regression analysis 122–3
transformations 87–9
   exponential growth model 102–4
   of input variables 87–8
   logit and inverse logit 109
   of output variables 89
treatments 72
trends 128–9, *130*

uncertainty 13, 16
   minimization 2
   in prediction 15
uniformity trials 58–*9*, *60*
unordered categorical data, graphical presentation 49
unpredictability viii

validity of experiments 8
variables, notational conventions 12
variance 52
  pooled 76
variation viii, 16, 17–20
  seasonality 129–30
  sources of 8
  stochastic 129
  trends 128–9, *130*
  in weather 127
variogram *146–7*, *150*
  directional 148, *149*
  isotropic *150*, *151*, *153*
variogram cloud 147–*8*, *149*

weather forecasting 127, 133–8
  lack of independent replication 127–8
weighted least squares criterion 88
wheat uniformity trial
  spatial correlogram 144, *145*
  variogram *146*–7
white noise model 129